NEW VANGUARD:

STURMARTILLERIE & PANZERJÄGER 1939–1945

WRITTEN BY BRYAN PERRET

COLOUR PLATES BY MIKE CHAPPELL AND MIKE BADROCKE

OSPREY
MILITARY

First published in Great Britain in 1999 by Osprey Publishing,
Elms Court, Chapel Way, Botley, Oxford OX2 9LP
Email: info@osprey-publishing.co.uk

ISBN 1 84176 004 8

Editor: Nikolai Bogdanovic
Design: Chris @ D4Design

Origination by Valhaven Ltd, Isleworth, UK
Printed through World Print Ltd, Hong Kong

99 00 01 02 03 10 9 8 7 6 5 4 3 2 1

Artist's note

Readers may care to note that the original centre cutaway painting from this book is available for private sale. All reproduction copyright whatsoever is retained by the Publishers. All enquiries should be addressed to:

Mike Badrocke,
37 Prospect Road, Southborough, Tunbridge Wells, Kent, TN4 0EN, UK

The Publishers regret that they can enter into no correspondence upon this matter.

TITLE PAGE
The Sturmhaubitze 42, armed with a 105mm howitzer fitted to the basic StuG III Ausf G. Production commenced in 1943 and thereafter accounted for one-eighth of assault gun production. (The Tank Museum)

STURMARTILLERIE & PANZERJÄGER 1939–1945

One of the two dozen or so StuG M assault guns to see service during the 1940 French campaign with Batteries 640, 659, 660 and 665. This is almost certainly an Ausf. B, employing the drive sprocket and idler of the narrow-tracked Ausf. A with spacer rings. The angled, spaced armour on the superstructure side is clearly visible, as is the broad V-shaped 'cut out' right of the gun, above the driver's compartment, leading back to the aperture of the gunner's sight. (RAC Tank Museum)

INTRODUCTION

All weapon systems are designed to overcome a particular aspect of the enemy's capacity to fight, and the majority are produced as a result of hard experience. Originally, the German assault gun was conceived to provide the infantry with the armoured support which had been lacking during the great battles of 1918, but changing conditions on the battlefields of the Second World War saw it develop into a powerful tank destroyer, although its infantry support role was never forgotten. The purpose-built tank destroyer had a much shorter history, and was produced to provide a direct answer to the heavy armour carried by certain British and Russian tanks.

If the much-glamorised Panzer divisions were the sword of the German Army, then the assault gun and tank destroyer units were its shield. In the last year of the war, the Panzers' once all-powerful grip on the battlefield began to fail, and more and more of the burden of anti-tank defence fell upon the assault gun and tank destroyer crews: it is largely due to their skill and professionalism that the massive Red Army did not advance even further into Europe.

THE CONCEPTION

During the First World War, only two effective means were found of breaking the deadlock of trench warfare on the Western Front. The first, employed by the British and French armies, involved the use of tanks, crushing their way through barbed wire aprons, crossing trenches and eliminating strongpoints by means of direct gunfire. In this context the tanks were simply carrying out what later became the standard role of the assault gun, the French even calling their own tanks *artillerie d'assaut*.

The second means, favoured by the German army, was infiltration on a massive scale following heavy bombardment. These operations were carried out by specially trained 'Storm Troops', who by-passed opposition and continued their advance into the enemy's rear areas without pause. However, some contact with the defence could not be avoided. Even where a front had been technically broken, knots of resistance varying in size from battalions down to single machine-guns continued to take their toll. Because of the shell-pitted, torn-up ground between the armies, the Storm Troops' horse-drawn supporting artillery could not be brought forward quickly enough to deal with these obstacles before heavy casualties had been incurred. Gradually, each successive offensive slowed to a standstill, until eventually the time came when the Storm Battalions had been bled so white as to be of no further use. They were Germany's best, and when they had gone the remainder of the army was good only for defensive warfare.

Another Ausf. B, photographed in the early stages of 'Barbarossa' with a load of infantry. The heavy 'unditching beams' are typical additions to the external stowage. The designation on the glacis suggests a battery HQ ('Stab') vehicle. (Bundesarchiv)

The second means merely provided a corollary for the first. Unless machine-gun posts, bunkers and strongpoints could be eliminated at the outset by direct gunfire, the infantry assault could succeed only at heavy cost. This was perfectly clear to the German General Staff in its analysis of the campaign after the war, but the provisions of the Treaty of Versailles, which specifically prohibited German acquisition or use of tracked AFVs, ensured that it would be many years before anything could be done about it.

In 1935 General Erich von Manstein[1] drafted a memorandum to the Chief of General Staff indicating that technical studies had shown the need for a self-propelled armoured gun to work under infantry control and give them support as required: he further suggested that each infantry division should contain an integral assault gun battalion consisting of three batteries each of six guns. Thanks to von Manstein's efforts, and the support of Generals von Fritsch and Beck, the project was approved: the artillery was given the task of designing the weapon system under the supervision of the General Staff's Technical Section 8, commanded by the then-Colonel Walter Model.

To save time it was decided to employ the already proven chassis and running gear of the Panzerkampfwagen III as a carriage. On this was placed a low, fixed superstructure with overhead cover and heavy frontal armour, mounting a limited-traverse L24 howitzer. The vehicle itself was constructed by Daimler Benz, while the gun was installed by Krupp. The completed prototype of this 'Sturmgeschütz III' was ready for trials on Kummersdorf Ranges early in 1937, and proved to be entirely satisfactory. It was hoped thereafter that by the autumn of 1939 each active infantry division would have its assault gun battalion, as would each reserve division sometime in 1940, although the number of guns in

[1] Von Manstein is regarded as the father of assault artillery by the German army, his memo containing mention of the words for the first time. It is interesting to note that even at this stage the diversion of Panzer units for infantry support in the British, French and Russian manner was considered to be a waste of resources.

each battery had been reduced to four. In the event, such optimism was quite unfounded.

First it had to be decided who was going to accept responsibility for the new weapon. Ought it to be the infantry, for whose benefit it had been developed? Or perhaps the Panzer troops, who were specialists in manning tracked vehicles? Or the artillery, who had been responsible for developing the original idea?

A conference was held between the respective Inspector Generals and their personal staffs. It was a conference which was to become memorable for its combination of histrionics and bathos. The Inspector General of Infantry began by explaining that his branch of service did not have the vehicle establishment that would be necessary to keep the assault guns supplied with fuel and ammunition; nor could it provide the technical services required to maintain the guns in the field. Rather than have to face these insuperable difficulties, he said he was prepared to let the whole idea drop.

The tank men wanted the project squashed at once on the grounds that it interfered with their own tank production programme, for which they wanted all available industrial capacity. In reply it was pointed out that the production of assault guns would leave the tanks free to get on with their own work; that if they were not produced, tanks would have to be diverted to support the infantry; and that the production figures for armoured carriages were rising so steadily that no manufacturing crisis was likely to arise. The Panzer officers were not impressed and remained stubborn and intractable. Tempers began to rise. Someone drew their attention to the fact that the assault gun's fixed superstructure permitted the installation of a larger calibre gun than that carried by the tanks, commenting with some asperity that their short-sighted views and lack of experience were combining to blind them to the meaning of this; that

A StuG III Ausf. E moving past a blazing Russian farm: the loader sits on the right-hand radio pannier, which became standard with this model. Note the Gothic 'D' at the left end of the rear hull plate. (Bundesarchiv)

assault guns would be able to knock out enemy armour at a range beyond their beloved tanks' capacity. At this point one Panzer officer completely lost control and, banging the table furiously, yelled that the conference 'had just passed sentence of death on the Panzer arm!' Somehow, it survived ...

It was now the turn of the Inspector General of Artillery to speak. He seems to have been a sleepy old gentleman, perhaps dreaming of the balmy days before 1914, when amid the thunder of hooves, his blue-coated gunners had swung their weapons into the battery line. He was aware that the subject under discussion concerned a new gun to support the infantry, and that it was causing a great deal of ill-feeling. In an effort to achieve a calmer atmosphere, he said that modern technology was all very well, but he felt that before any serious decision was taken trials should be held to decide whether the new support gun would not be better horse-drawn, in the manner of the First World War. While jaws gaped in astonishment, his embarrassed personal staff gathered round to explain von Manstein's ideas on tactical employment, as well as the nature of the weapon itself. It took a little time to bring the general up to date, but once he had arrived he began to warm to the concept, which he agreed was best handled by artillerymen, much to the relief of everyone present. (Thereafter, it was very much an 'in' joke among senior officers when referring to assault guns to describe them as 'Horse-Drawns'. The crews called them 'Snouts', a corruption of Geschütz into Geschnauze.)

Thus, after a period of being everyone's baby and nobody's child, the assault gun returned to the control of artillery. The Artillery School at Jütebog was detailed by the Inspector General to establish basic training facilities and a tactical school for assault artillery. In the autumn of 1937 an Experimental Battery was set up by the 7th Motorised Artillery Demonstration Regiment, and this carried out a variety of exercises throughout the following winter. Once the results of these had been evaluated, the Experimental Battery spent a year carrying out combined trials with the Infantry Demonstration Regiment at Doberitz, during which tactical principles were established for the mutual benefit of both arms.

These trials were carried out by the prototype vehicles, supplemented by PzKpfw III chassis mounting a dummy superstructure and gun. For security reasons the assault gun was referred to throughout as a 'self-propelled 37mm anti-tank gun'. During gunnery trials the assault gun crews did rather better than their Panzer counterparts using the same weapon fitted to the PzKpfw IV, being quicker onto the target and using less ammunition to destroy it. This was particularly satisfactory for the protagonists of the new weapon system, who had been forced to develop their techniques without assistance from the tank men.

A small number of assault guns were sent to Tunisia shortly before the collapse, and this photo shows an Ausf. F, with L48 gun, in that theatre. The paint scheme seems to be overall light desert yellow, with applied scrub foliage. This may be a gun of the incomplete 'Hermann Göring' Division. (Bundesarchiv)

An infantry section ride an assault gun in Russia. Commanders were supposed to ensure that only complete tactical units rode any one gun, to avoid confusion on the operation start line. (Bundesarchiv)

Unfortunately, the protracted nature of the troop and gunnery trials, which in themselves had proved entirely satisfactory, combined with other factors to delay the series production of assault guns, so that when war broke out not one single battery was available for service use.

Two concepts of anti-tank defence; a StuG III Ausf. G, with 'Saukopf' mantlet and skirt armour, for distant targets, and the infantryman's hand-held Panzerfaust for close-quarter fighting. (Bundesarchiv)

THE STUG III ASSAULT GUN

The StuG III assault gun holds a very special place in the history of German armour, if only because it was the only tracked weapon system to serve throughout the Second World War in the same basic format. This consisted of only two compartments; the rear containing the 300hp Maybach engine, and forward of it the squat, angular superstructure housing the gun and crew. The commander was located at the left rear of the fighting compartment directly behind the gunner, who was in turn seated behind the driver. To the right of the gun were the loader/operator and his ammunition racks containing 44 rounds. The radio, a 10-watt UKW, was carried in a pannier on the left, although a second radio was usually carried on the right in command vehicles.

The absence of turret meant a considerable saving in weight which was translated into 50mm frontal armour, far thicker than that carried by contemporary German tanks; it also meant a low silhouette, the vehicle's overall height being only 6ft 4in. On the early models the vertical side walls of the superstructure were further protected by angled 9mm plates which provided a form of spaced armour. Weighing 21½ tons, the StuG III could achieve a speed of 25mph, comparable with most medium tanks of the period.

The assault gun's most obvious fault was the limited traverse available to the gunner, but in the circumstances this was unavoidable. Other faults included the gunner's vulnerable periscopic sight and the fact that the commander was forced to observe through his open roof hatch, using periscopic binoculars. The early models also lacked a machine-gun for close defence.

Nonetheless, through regular modification the StuG III kept pace with battlefield technology.The major improvements are listed below, together with their period of introduction.

The Ausf. G saw the introduction of a commander's cupola. The periscopic binoculars visible here have extension tubes fitted, to keep rain and dirt off the lenses. Note the interesting angular camouflage painting of the skirts, the forward plate of which seems to be a replacement from another gun. The external stowage has been arranged to leave the engine deck hatches clear. While it lasted, the cabin trunk at the back must have been a very useful acquisition! (Bundesarchiv)

Ausf. B

Summer 1940. Improved transmission fitted. Width of tracks increased to 40cm, necessitating new drive sprockets and rear idlers, although some vehicles retained their old sprockets and idlers, modified by the addition of spacer rings to compensate for the extra width. The new sprocket was slightly dished and contained six openings as opposed to the eight holes which distinguished the older version; the new idler was an eight-spoked wheel, whereas that carried by the Ausf. A had been solid.

Ausf. C and D

Early 1941. Superstructure redesigned to eliminate the gunner's sighting V-slot from the roof, which had proved to be a weakness on Ausf. A and B: the gunner's sight now protruded through the roof. The Ausf. C and D were externally identical, the latter having various internal modifications.

Ausf. E

Autumn 1941. The 9mm external superstructure side plates carried on Ausf. A–D were dispensed with. Instead of being angled, the superstructure side walls were now of vertical 30mm plate. The left radio pannier was extended forward and a right–hand pannier fitted as standard, being used for ammunition stowage when no second radio was carried.

Ausf. F

Early 1942. The first encounters with the Russian KV and T-34 tanks had shown them to be not only heavily armoured but also well armed. To counter this the assault gun was re-equipped with a longer (L43) 75mm gun which entered the vehicle through a prominent 30mm block

Swirling smoke and dust make a dramatic background for this aerial photo of StuG IIIs going into action with Panzer-Grenadiers in half-tracks. The camouflage painting on these guns resembles giant chicken-mesh – regular hexagons of brown or green lines on the ochre finish. The terrain suggests the southern Russian front. (Bundesarchiv)

mantlet, while additional 30mm plates were fixed to the front armour. Additional protection was sometimes obtained by pouring concrete infill into the hollows at the front of the superstructure roof. Other modifications included the installation of a ventilator in the centre of the roof, the provision of a close-defence machine-gun and shield mounted in front of the loader's hatch, and re-arrangement of the engine deck hatches.

The Ausf. F had only been in production a short time when it was decided to fit the more powerful L48 gun in place of the L43. This modification was carried out retrospectively to existing Ausf. Fs as well as being incorporated into the manufacture of new vehicles, which were known as Ausf. F/8s.

The new gun provided a very welcome antitank capacity, for which the slight reduction in the amount of ammunition carried was a small price to pay. Officially 42 rounds could be stowed, but it is said that with careful layering it was possible to carry as many as 120 rounds – presumably with a crew of midgets!

Ausf. G

Early 1943. 80mm frontal armour became standard with this model, on which the superstructure had been widened over the tracks, incorporating the radio pannier space. The commander was provided with a cupola containing eight independently retractable episcopes: within the cupola was a split hatch, and by opening the smaller frontal section it was possible for the commander to use his periscopic binoculars under cover of the closed rear section. Various minor modifications included the removal of the ventilator from the roof to the rear wall of the fighting compartment, and re-alignment of the loader's split hatch to open fore and aft.

While some early Ausf. Gs retained the block mantlet carried by the Ausf. F/8, the majority were fitted with a more satisfactory cast type known as a 'Pig's Head' from its shape. As a protection against hollow-charge ammunition large plates known as 'skirts' (*schuertzen*) were hung along the side of the vehicle, which also received a coating of anti-magnetic Zimmerit paste, designed to prevent the attachment of mines and other devices in close combat. Some vehicles carried a battery of smoke bomb dischargers, located three on either side of the superstructure, but this was by no means universal.

This head-on view of a very late production StuG III Ausf. G shows the superstructure extending out over the tracks; the toothed quick-release mounting rail for skirt plates; a liberal coating of Zimmerit plaster; concrete infill poured into the roof depressions; and what appears to be the remote-control roof machine-gun fitted to late models. (RAC Tank Museum)

Sturmhaubitze 42

Concentration upon increasing the assault gun's anti-tank capacity tended to militate against the primary role for which the weapon system had been designed, i.e. infantry support by direct gunfire. The disappearance of the short 75mm L24 models was in some measure compensated for by the production of the Assault Howitzer 42, which carried a 105mm howitzer fitted to the basic StuG III Ausf. G. Although the prototype appeared in 1942 (employing an Ausf. F superstructure) quantity production did not begin until the following year, and thereafter accounted for only approximately one-eighth of assault guns manufactured. (A total of 10,500 assault guns on the PzKpfw III chassis were built throughout the war, the majority by Alkett of Berlin.)

Assault guns based on other chassis were produced in much smaller numbers, and the main types were as detailed below.

Excellent view of a Brummbär in Italy, 1944 – see colour plate G3. The operator is extending his aerial in preparation for netting-in. (Bundesarchiv)

StuG IV

This vehicle was a hybrid designed to supplement standard assault gun manufacture. It consisted of an Ausf. G superstructure mounted on a PzKpfw IV chassis, the armament being the L48 75mm gun. It first entered service in mid-1943 and by the end of the war 632 had been built. Its PzKpfw IV running gear and prominent driving compartment make it easily recognisable.

'Brummbär'

Street fighting in the cities of Russia revealed that the 75mm gun lacked the necessary punch to deal with well-constructed buildings. The 150mm L12 howitzer was considered to be the smallest gun with muscle enough for the job, and in October 1942 work began on producing a heavy assault gun specially designed for close combat in built-up areas.

The chassis chosen was that of the PzKpfw IV, on which was mounted a heavy, angled superstructure, the gun being housed in a large ball mounting in the 100mm front plate. Weighing 28 tons, the equipment tended to overload the chassis, causing transmission problems; but otherwise the vehicle proved to be satisfactory, and entered service in April 1943 under the nebulous title of Sturmpanzer IV. This was soon dropped in favour of the more descriptive 'Grizzly Bear'.

A total of 313 Grizzly Bears were built, running to several models which incorporated minor improvements. They served with the Heavy Infantry Gun companies of Panzer-Grenadier regiments, and also in 45-strong Assault Battalions which were at the disposal of senior commanders.

An interesting upper view of the Sturmmorser Tiger, showing details of the weapon's ball-mounting and mantlet, rocket loading hatch in the roof with guide rails below, and hoist. (The Tank Museum)

Sturmtiger

This represented the Grizzly Bear idea carried to extremes. While possessing a similar layout, this vehicle employed a 380mm rocket launcher to fire a 761 lb spin-stabilized missile up to 6,000 yards, although normally engagements would have been carried out at much closer range. The missiles were of two types, one carrying a normal high explosive payload and the other a shaped charge for use against concrete. A small crane was provided to assist the loading of the missiles into the launcher, which was fitted with a heavy counter-weight.

Built on a Tiger Ausf. E tank chassis and protected by 150mm armour, the 68-ton Assault Tiger was something of a misfit. By the time it appeared in 1944 the German Army was wholly engaged in defensive battles, and its potential role had long since vanished. Only a handful of these monolithic vehicles were produced.

An abandoned Brummbär with skirt armour: note the late driver's vision block, identical to that used on the Tiger Ausf. E, and details of the ball-mounted 150mm howitzer. (RAC Tank Museum)

MANNING AND ORGANISATION

Assault artillerymen were all volunteers and were considered to be the élite of their service. As well as gaining numerous awards of the Iron Cross First and Second Class, the Assault Artillery branch won no less than 325 German Crosses in Gold and 140 Knight's Crosses, 14 holders of the latter being further awarded Oakleaves to their decorations. It goes almost without saying that assault artillerymen were extremely proud of their calling and their *esprit de corps* was always high. Many units chose distinctive identification symbols, which were painted on their vehicles apparently at the discretion of their current commanding officers.

By 1940 the number of guns in the assault battery had been increased again from four to six, and the success of the few units engaged in France led to considerable enthusiasm for the assault gun idea. Full-scale production was commenced and the Artillery School at Jütebog began training the first Abteilungen, or battalions.

Initially the pace was slow, only two battalions being produced every three months, but this later accelerated to three battalions every two months. In 1943 the Assault Artillery established its own school at Burg near Magdeburg, and the flow of trained units and replacement personnel increased rapidly. Eventually some 70 battalions were produced, excluding units specially trained for the tank destroyer role, or sent to reinforce understrength Panzer Divisions.

The new battalions consisted of a Headquarters Battery (containing the commanding officer's assault gun, the battalion transport echelon, recovery, workshop and medical services) plus three Assault Batteries, each of three two-gun troops, battery echelon and fitters. The internal organisation of the battery was later strengthened by adding one gun to each troop and by providing the battery commander with his own gun in place of the armoured command vehicle from which he had originally exercised control, thus giving a total Abteilung strength of 31 fighting vehicles.

In 1943 the Sturmgeschütz Abteilung title was changed to Sturmgeschütz Brigade, largely in the hope that this would suggest to the enemy that

A Sturmtiger, showing missile derrick, and prominent muzzle counter-weight on the rocket launcher. (RAC Tank Museum)

A fine side view of Germany's first tracked tank destroyer, the Panzerjäger I. This clearly shows the structure of the vehicle. (Bundesarchiv)

larger assault gun formations were being employed than was the case.

In 1944 the first of the Sturmartillerie Brigades were formed. More than a slight change in nomenclature was involved. These units had a much larger establishment, comprising three L48 guns in Brigade HQ, two L48 guns in each Battery HQ, plus two troops of four L48 guns and one troop of four StuH 42 howitzers per battery; a total of 45 fighting vehicles. For defence against infantry anti-tank weapons in close country or during street fighting, the establishment was further swelled by the incorporation of a Grenadier Escort Battery with a nominal strength of two officers and 196 men, mostly equipped with assault rifles, and which included a small pioneer troop. It had been hoped to raise all Assault Gun brigades to Assault Artillery Brigade strength, but this was not possible.

Meanwhile, at the beginning of 1943, the German tank industry had fallen into complete chaos, and manufacture of assault guns was steadily out-stripping that of battle tanks, which were more difficult and expensive to construct. Production of the PzKpfw III had been discontinued in 1942, save for the chassis which were required for assault guns: now there was a suicidal but apparently serious suggestion that the PzKpfw IV, the mainstay of the Panzer Divisions, might be cancelled as well in order to concentrate on Tiger and Panther production. A dangerously high percentage of PzKpfw IV chassis had already been allocated for self-propelled artillery, tank destroyers and assault guns: Tigers were coming off the assembly line at the rate of only 25 per month; and the Panther had not yet entered full production.

To restore some semblance of order Hitler was compelled to recall the brilliant

Another view of the Panzerjäger I, showing it in action in the outskirts of Stalingrad in the late summer of 1942. Protection for the crew of the 47mm gun was minimal. (RAC Tank Museum)

The Marder II mounted the 75mm Pak 40/2 anti-tank gun on the chassis of the PzKpfw II Ausf. A, B, C and F. (RAC Tank Museum)

but temporarily disgraced Col.Gen. Heinz Guderian and appoint him Inspector General of Armoured Troops with a very wide brief which included the Waffen-SS and Luftwaffe, but not the assault artillery. When Guderian attended a senior officers' conference on 9 March 1943 he attempted to rectify this omission, but his suggestion that all assault artillery should become subordinate to him was howled down, apparently to Hitler's delight. The reason for this was not – as Gen. Schmundt, Hitler's principal adjutant, so fatuously claimed – that the assault artillery was the only branch which enabled gunners to win the Knight's Cross; rather it was a matter of changing battlefield technology.

Following the introduction of hollow-charge ammunition which could penetrate any thickness of tank armour, and the proliferation of such devices among the infantry, Hitler had already intuitively grasped that the tank's era of total domination was at an end. His thoughts were now concentrated on anti-tank defence, and since the assault guns performed this aspect of their work most efficiently and were not suffering as a result of the production crisis, he saw no reason why they should be subject to interference. It was a view that many senior officers supported.

Guderian had a second point to make concerning assault artillery, and one which was obviously valid. 'Anti-tank defence will devolve more and more on the assault guns, since all our other anti-tank weapons are becoming increasingly ineffective against the new enemy equipment ... All divisions on the main battle fronts, therefore, need to be supplied with a certain complement of these weapons; the secondary fronts will have to make do with a higher command reserve of assault guns, while the divisions are for the time being equipped with self-propelled anti-tank guns. In order to economise on personnel and material, a gradual amalgamation of the assault gun battalions and tank destroyer battalions is necessary.'

Firing over unfrozen snow largely negated the value of smokeless propellants and gave away a gun's position at once. (Bundesarchiv)

This was accepted, after what Guderian describes in his book *Panzer Leader* as a 'lively discussion'. In the meantime, the tank element of most Panzer divisions had fallen to a dangerously low level, and out of sheer necessity several had to be urgently reinforced with assault guns, which could obviously provide heavy fire support in the forward zone, but which could not be expected to carry out the same functions as the tanks themselves.

In general, assault artillery was allocated as and where it was needed. Very few divisions had an integral assault gun batallion, but the expanding 'Großdeutschland' formation was an exception, as were the SS-Panzer-Divisions 'Leibstandarte Adolf Hitler' and 'Das Reich', and the Luftwaffe Panzer Division 'Herman Göring'.

TACTICS

The following extracts have been taken from what might be termed The Assault Gun Commander's Handbook, issued in 1944.

'General Principles

You have the task of providing immediate direct support for the infantry in all situations in the forward zone of action, beating down and suppressing the fire of the enemy's heavy weapons with that of your own.

The assault gun combines firepower with mobility and shock action. Its protected weapon system and immediate direct fire capability enables it to accompany the infantry anywhere on the battlefield, providing them with close physical and moral support. Assault guns are employed according to artillery principles and should be regarded as first line artillery. Concentration is a special feature of their operations; abandonment of this principle leads to unnecessary loss.

In every action the destruction of the enemy's tanks is a consideration of the utmost importance. Nonetheless, you must not permit your assault guns to be employed solely as tank destroyers.

In the counter-attack role assault guns are suitable for employment as a shock reserve against enemy tank attacks or penetrations of our own lines. As the spearhead of the advance or pursuit they can quickly overcome points of resistance.

When breaking off action and disengaging, assault guns, by virtue of their mobility and firepower, are the mainstay of the rearguard.

A temporary application of gunfire for specific bombardment in the forward battle zone is permissible, provided other artillery units are unable to fulfil this task, and provided it does not interfere with the

The much-photographed Marder II nicknamed 'Coal Thief' – the German music hall equivalent of the British 'spiv'. There are already more than 20 kill rings round the gun barrel. This vehicle has been identified by one German writer as serving with 29.Inf. Div. (Bundesarchiv)

The captured Russian 76.2mm anti-tank gun was rechambered to take German 75mm rounds, and mounted on the chassis of the obsolete Czech 38t to produce the Marder III. In May 1942, 117 of these vehicles were sent to North Africa to counter the menace of the heavily armoured Matilda II. (Bundesarchiv)

assault guns' primary mission. In such situations the assault howitzers will undertake the fire task.

Assault guns are unsuitable for employment in static situations. Their only suitable and sensible function is close co-operation during infantry, Panzer-Grenadier or tank operations.

Employment

For operations you will be placed under the command of an infantry or Panzer-Grenadier division, and occasionally a tank division; employment under the command of units smaller than regiment is the exception.

The formation commander must be made aware of the assault gun's characteristics, so that the maximum benefit can be obtained from the combined operation. The concentrated fire and shock effect of the whole brigade working on a narrow front produces the best results.

The tactical unit is the brigade. To parcel out guns in troops or singly impairs the mass fire effect and assists the enemy's defence. Support of the infantry by single troops must, therefore, be limited to exceptional cases, and the troop promptly returned to brigade control at the conclusion of the operation. Employment of single guns in the forward zone should never take place, since assault guns are designed to provide mutual assistance for each other in circumstances of tactical or technical difficulty.

The more surprise the assault guns can achieve, the better will be the result. The approach march and final preparations are best made by night, the noise muffled by loudspeakers or artillery fire. Assault guns will not take part in the preliminary bombardment.

The brigade commander will arrange for replenishment and technical assistance facilities to be available in the forward zone, and assess the correct moment for the guns to be withdrawn to this rally point.

Co-operation with infantry

Infantry must make full and immediate use of the assault guns' fire support. Fire and movement between infantry and guns must be mutually agreed. The best results are obtained when the infantry and guns adopt a loose formation. As the assault guns will attract most of the enemy's fire, the infantry should not crowd together behind them, as this will result in heavy casualties; instead, the infantry should advance by a series of carefully timed rushes.

Co-operation between assault guns and the infantry's heavy weapons must be agreed beforehand, particularly in cases where the former are at risk from the enemy's anti-tank guns.

In open country, the assault guns will lead the advance; where the country is closer, the infantry will take the lead. [In addition to the difficulties posed by woodland, assault guns were blinded by any tall standing crop; the document makes specific mention *inter alia* of sunflower fields.] In the latter case it is the responsibility of the infantry to warn the guns of any obstacles in their path, such as minefields, bad

A Marder III coming out of action with wounded aboard. (US National Archives)

going and anti-tank positions. Mutual identification of targets can be achieved by means of tracer ammunition, flare pistols, indicative shooting and hand signals. The assault gun and infantry commanders should remain in the closest contact throughout the action.

Co-operation with field artillery

As the fire of field artillery is supplementary to that of assault guns, strict co-ordination of their respective fire plans is essential.

By means of their radios, the assault guns are responsible for ensuring that supporting artillery fire is put to its best use; keeping the field artillery in contact with the spearhead of the attack; and providing early warning of defensive fire tasks that might be required.

The field artillery are responsible for neutralising the enemy's artillery and fire control posts; preparing the way for the attack and protecting its flanks; and providing protection for the recovery of assault guns which have become casualties.

The co-operation of assault guns with the Artillery Fire Controller is, however, a supplementary task and must not detract from the principal mission of providing direct fire support for the infantry.

Co-operation with pioneers

Close co-operation between assault guns and pioneers prevents casualties and losses. Under cover of the assault guns' fire the pioneers will create minefield gaps, fill in anti-tank ditches and strengthen damaged bridges. During an attack against prepared positions, each assault gun battery will be allotted pioneers for the duration of the operation.

Assault guns are outstandingly suitable for the support of pioneers in all manner of battle situations.

Co-operation with Luftwaffe

The involvement of the Luftwaffe in the ground battle will benefit the assault guns by causing confusion among the enemy and by raising dust and smoke clouds. In order to advise the support squadrons as to the position of the front line, assault guns will put out orange smoke markers, and by firing smoke or tracer ammunition they can advise the aircraft of suitable targets.

For co-operation with the Luftwaffe the assault gun brigade is equipped with the Air Co-operation Radio Set FU7.

Tasks of escort grenadiers

The first duty of the grenadiers is the protection of the assault guns in all phases of the battle. Escort grenadiers are not assault troops; they are neither armed nor able to perform this role, having a specialised task to perform.

As well as providing a close escort for the guns, the grenadiers will; indicate targets by tracer fire or hand signals; warn the assault guns of anti-tank positions, minefields and other obstacles; secure the route to the operational start line; and protect the guns in their harbour area.

Once allocated to a particular gun, escort grenadiers will remain with it throughout the action. If they are in danger of falling behind, they will attract the vehicle commander's attention by firing their assault rifles into

A Marder III photographed on the open steppe, autumn 1942. Apart from the very limited space available for stowing the crew's kit and the various 'extras' always acquired on campaign, the vehicle was seriously handicapped in being able to accommodate only 30 rounds of main armament ammunition. This crew appear to wear the reed-green vehicle denims, with a black Panzer sidecap with pink soutache. (Bundesarchiv)

the ground ahead of the vehicle, and do their utmost to regain contact.

Should the establishment of the Brigade not include an Escort Grenadier Battery, the infantry will be detailed to supply a sufficient number of men to perform the task.

Enemy tanks

Should assault guns encounter tanks, they are to be engaged. Measures to be taken include the construction of a fire-front and engaging the flanks of the enemy formation.

Late-model Marder IIIs were armed with the German 75mm Pak 40/3, with the engine relocated 'amidships' and the fighting compartment at the rear. Note the small rectangular hole cut in the armour right of the gun, for the gunner's telescope; and the travelling crutch for the gun. To free the armament the gunner simply elevated the barrel, and the crutch fell away. (RAC Tank Museum)

Where the enemy attack is accompanied by infantry, these will be engaged by the assault howitzers alone, while the remainder of the guns concentrate on the tanks; this will have the effect of separating the enemy infantry from their armour. For this type of action an internal troop organisation of one assault howitzer and two assault guns is particularly suitable.

Terrain and weather

The shape of the ground, the nature of its covering and the type of going all influence assault gun operations. Full utilisation of terrain increases weapon capability, decreases losses and saves wear on equipment.

In winter a snow depth of 50cm will prevent the full use of assault guns, and special precautions must be taken in view of the effect of extreme cold on their engines. Severe cold will also affect ammunition, so that ricochets and bursting are both likely: the howitzer can maintain its efficiency by employing a double charge. Service with assault guns is arduous and a real danger of freezing to death exists; therefore, measures against the cold require particular attention and dedication.'

ACTIVE SERVICE

No assault gun units were available for service during the 1939 Polish campaign. However, by the summer of 1940 six batteries had been formed, four of which saw service in France (No. 640, 659, 660 and 665), one fighting in support of the then-infantry regiment 'Großdeutschland'. During the 1941 Balkan campaigns against Yugoslavia and Greece three battalions were employed, and this figure was doubled for the invasion of Russia, Army Groups North, Centre and South each having the support of two battalions. Thereafter, as soon as they became available, further battalions were despatched to the Eastern Front in rapid succession.

The subsequent history of the arm follows the triumphs and tragedies of the war with Russia, but assault gun units also saw active service in Finland, Italy, Normandy and on the Western Front. Their involvement in North Africa was limited to a handful of vehicles despatched to Tunisia just before the final collapse: during the desert phase of the campaign there had been little call for assault guns in what was essentially a tank battle.

To record even the major events in the Assault Artillery's dramatic

A first-class gun platform, the Nashorn was poorly protected, and its great size made concealment difficult. A heavy, soft mottle of brown and green patterns the basic ochre of this vehicle: a tactical marking (a 'T' rising from the oval 'tracked' symbol) can just be made out below and right of the open doors. Unfortunately the right-hand door covers a large, shield-based unit emblem of which the edge is just visible. (Bundesarchiv)

anti-tank gun which fired a 12.6lb projectile at a muzzle velocity of 1,800ft/sec, and carried a crew of four. The equipment weighed 11 tons and had a top speed of 25mph. A total of 531 of these conversions were made during 1942 and 1943, and saw service in the anti-tank units of infantry and Panzer divisions.

The chassis of the PzKpfw II Ausf. D and E were used as the basis for another equipment also known, confusingly, as Marder II, although in appearance the two vehicles were quite different: the Ausf. D and E versions had a high, box-like superstructure further aft and were equipped with a captured Russian Model 36 76.2mm gun re-chambered to take the German 75mm round. This weapon had a muzzle velocity of 2,430ft/sec, later increased to 3,250ft/sec with improved ammunition, and proved capable of defeating Russian armour. 185 of these vehicles were built concurrently with their namesakes, and were similarly employed. Occasionally guns were transposed between the respective chassis, according to availability: sometimes muzzle brakes were fitted, sometimes not.

The Russian gun was also used on the early models of the Marder III, which was based on the chassis of the Czech 38t tank, and which also employed a high box superstructure for the gun mounting. In this form a total of 344 vehicles were built, of which 117 were sent to North Africa in May 1942 to combat the Matilda menace. They were so successful that the British thought, prematurely, that they were dealing with a mechanised version of the dreaded '88'.

Later models of the Marder III mounted the German 75mm L46 Pak 40/3 gun, which gave a similar performance to the Russian weapon. This appeared in two forms, the first having the fighting compartment well forward, while on the second the engine had been moved amidships and the fighting compartment installed at the rear. Most of the 1,577 75mm Marder IIIs built were employed in Russia in the same capacity as the Marder IIs. They carried a crew of four, weighed 10.5 tons and had a maximum speed of 26mph.

A number of captured French chassis were converted for tank destroyer use in 1943. Several were fitted with obsolete 37mm and 47mm anti-tank guns, but the most important result was obtained by mounting the German 75mm Pak 40/1 gun on the Lorraine tracked carrier, the combined equipment being known as Marder I.[1] The 75mm gun was also fitted to the Hotchkiss and FCM tank chassis in small numbers. In general the French conversions were regarded as second-line equipment

[1] This brought the number of 'Marders' in service to six, all of different appearance, leaving contemporary recognition experts confounded; in some measure this confusion persists to this day, and both guns and chassis must be specified if the name is to be used with any clarity.

and were employed mainly in occupied countries, where they served in the training or counter-invasion roles.

Last and largest of the interim tank destroyers was the Nashorn ('Rhinoceros'), which was introduced in 1943. This vehicle solved the difficulty of providing mobility for the 88mm anti-tank gun on the Eastern Front, the L71 Pak 43/1 model being installed in a fighting compartment which occupied most of the front-engined PzKpfw IV chassis, which incorporated the transmission and final drive of the PzKpfw III. The Nashorn was issued to 30-strong heavy tank destroyer battalions, which were committed under army control as the situation required. While combining mobility with heavy firepower, they were poorly protected for a front-line weapon, and their great height (9ft 7½ in) was a disadvantage when it came to the selection of concealed fire positions. They weighed 24 tons, had a crew of five, and could travel at 26mph. Their name had been conferred by Hitler himself, who apparently considered that the original title of 'Hornet' lacked ferocity.

Subsequent equipments may be grouped as 'second generation' tank destroyers, and they are detailed in the paragraphs that follow.

The 88mm Pak 43/2 was used by the first fully enclosed tank destroyer, the Elefant, sometimes also referred to as the Tiger (P) and sometimes as Ferdinand, after its designer, Dr Ferdinand Porsche. The chassis was, in fact, that of the unsuccessful Porsche design for the Tiger tank, and the layout consisted of a forward compartment for the driver and radio operator, the engine compartment amidships, and an enclosed fighting compartment at the rear. The frontal armour was 200mm thick and the vehicle weighed 67 tons, being powered by two Maybach 320hp engines, which produced a maximum speed of 12mph.

Of the 90 produced, 76 entered service in July 1943 with two 38-strong heavy tank destroyer battalions, numbered 653 and 654, at the battle of Kursk. Here they were employed in a maverick role, that of breakthrough tanks, for which they had not been intended. While they found no difficulty in smashing the opposing armour, their supporting infantry were soon left behind. Having no close-defence machine-gun and being all but blind into the bargain, they quickly fell prey to Russian tank-hunting teams which surrounded them and closed in with flamethrowers, Molotov cocktails and explosive devices. Needless to say, both battalions suffered severely.

Those Elefants which survived Kursk were fitted with a hull machine-gun and sent to Italy. Their great weight and height and low speed meant that they lacked the mobility of a true tank destroyer, but they performed very well in the semi-static conditions of mountain warfare, their thick frontal armour proving quite impervious to Allied anti-tank guns.

The Jagdpanzer IV was, as its name implies, based on the chassis of the PzKpfw IV, but was also commonly known as 'Guderian's Duck'. Its layout followed that of the conventional assault gun, with the fighting compartment forward.

The superstructure consisted of well-angled armour plate, extended to the vehicle's rear,

The 67-ton Elefant had a disastrous debut at Kursk, where it fell victim to infantry tank-hunting squads against which it had no defence. Most survivors of 'Zitadelle' were sent to Italy, where they were valuable in the semi-static defensive conditions. This one is evidently a mine casualty. Note the solid 100mm armour slab bolted to the front plate, which was itself 100mm thick. The rudimentary cupola and the machine-gun mounted through the frontal armour were improvements made as a result of the losses at Kursk. (Bundesarchiv)

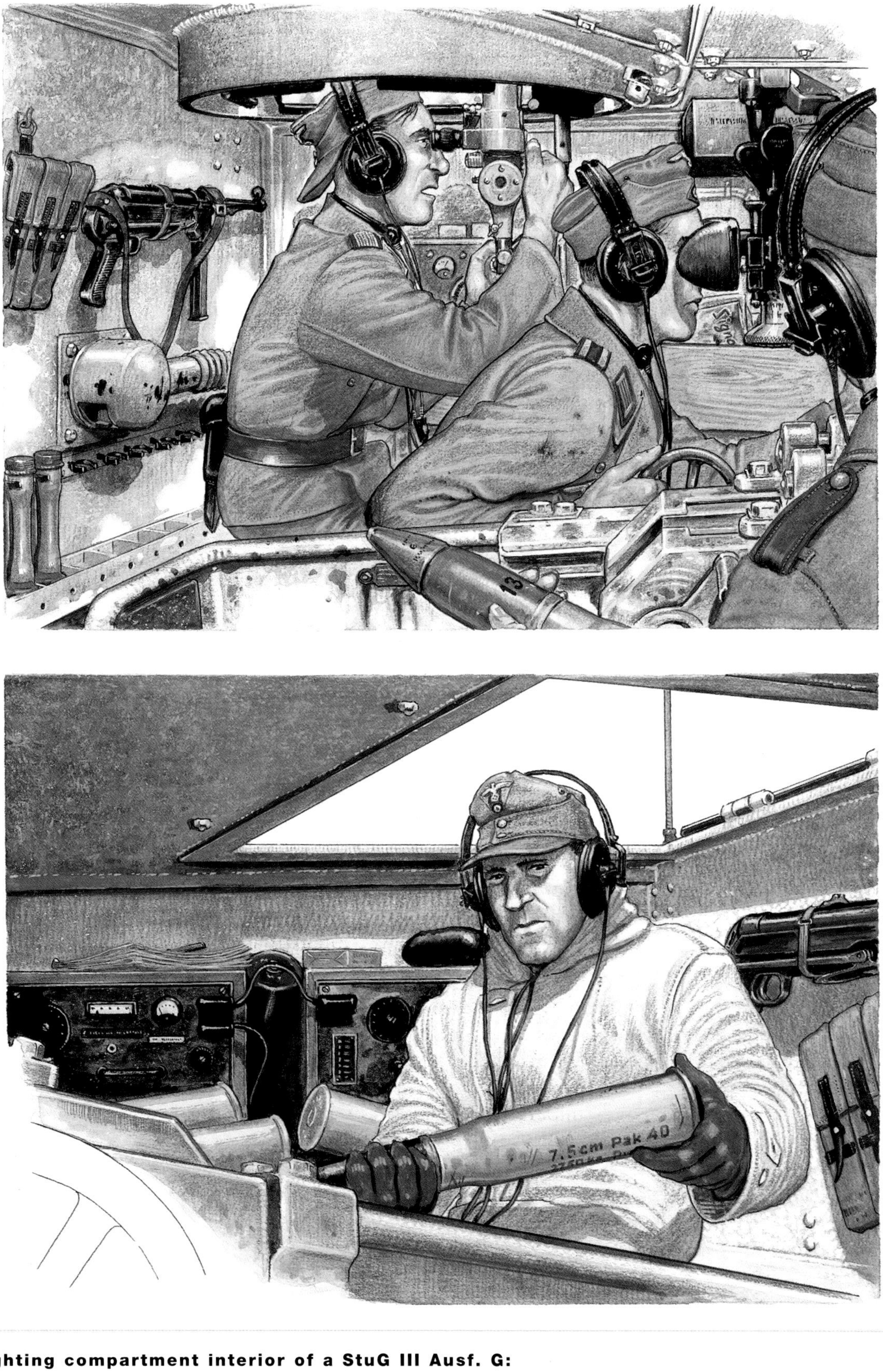

A

Fighting compartment interior of a StuG III Ausf. G:
the top illustration shows the left, and the bottom the right hand side.

1. StuG III Ausf. B, 2 Batterie, StuG Abteilung 192; Russia, August 1941.

2. StuG III Ausf. B, 3 Batterie, StuG Abteilung 203; Smolensk, Russia, July 1941.

**1. StuG III Ausf. G,
2.SS-Panzer-Division 'Das Reich';
Kursk, Russia, July 1943.**

**2. StuG III Ausf. G,
16.SS-Panzer-Grenadier-
Division 'Reichsführer';
Italy, January 1944.**

JAGDPANZER IV (SDKFZ 162)

KEY

1 Cast drive sprocket
2 Quick release headlight with blackout cover
3 Marking for Panzer Lehr Division
4 60mm nose armour
5 Steering brake inspection hatch with air intake
6 Transmission inspection hatch
7 Armoured mantlet
8 7.5cm PaK 39 L/48
9 Muzzle brake removed
10 Towing eye (extension of side armour)
11 Track link tool
12 Fire extinguisher
13 Right side MG ball mount with armoured cover
14 Front armour plate 60mm
15 Gun mount armour
16 Roof armour
17 Right side sponson front ammunition rack
18 Sliding armoured cover over gun sight opening
19 Loader's periscope
20 Sfl ZF 1 periscopic gun sight
21 Loader's escape hatch
22 Right side sponson rear ammunition rack
23 Duct pipe to extractor fan in read plate
24 Radio antenna (2m rod)
25 FU 5 radio sets
26 Blanked off opening for Nahverteidigungswaffe (close-in defence weapon)
27 Gun cleaning rods
28 Spanner for locking track idler wheel adjustment
29 Starting handle
30 Air outlet
31 C hooks for use with towing hawser
32 Spade
33 Spare radio antenna rod
34 Axe
35 Wire-cutters
36 Spare track links
37 Air intake
38 Spare road wheels
39 Exhaust silencer
40 Jack
41 Engine compartment side 5mm spaced armour
42 Air intakes on left of motor compartment
43 Radiators
44 Tool for adjusting track idler wheel
45 Fabricated idler wheel
46 Tracks kgs 61/400/120
47 Steel return roller
48 Road wheel, tyre size 470 x 90-359
49 Quarter eliptical spring mounting
50 Blanked off opening for FU 8 radio antenna on command vehicles
51 Commander's escape hatch with rotating periscope
52 Left side periscope
53 Left side sponson rear ammunition rack
54 Battery box
55 Folding racks for ammunition on hull wall
56 Recoil guard
57 Seat for gunner
58 Floor escape hatch
59 Traverse hand wheel
60 Elevation hand wheel
61 Bump stop
62 Fuel filler pipe
63 Armoured fuel tank
64 SSG 76 gearbox and clutch
65 7.92mm MG42 machine-gun in optional gunners mount
66 Driver's seat
67 Left side MG ball-mount with armoured cover (on early vehicles only)
68 Driver's periscopes

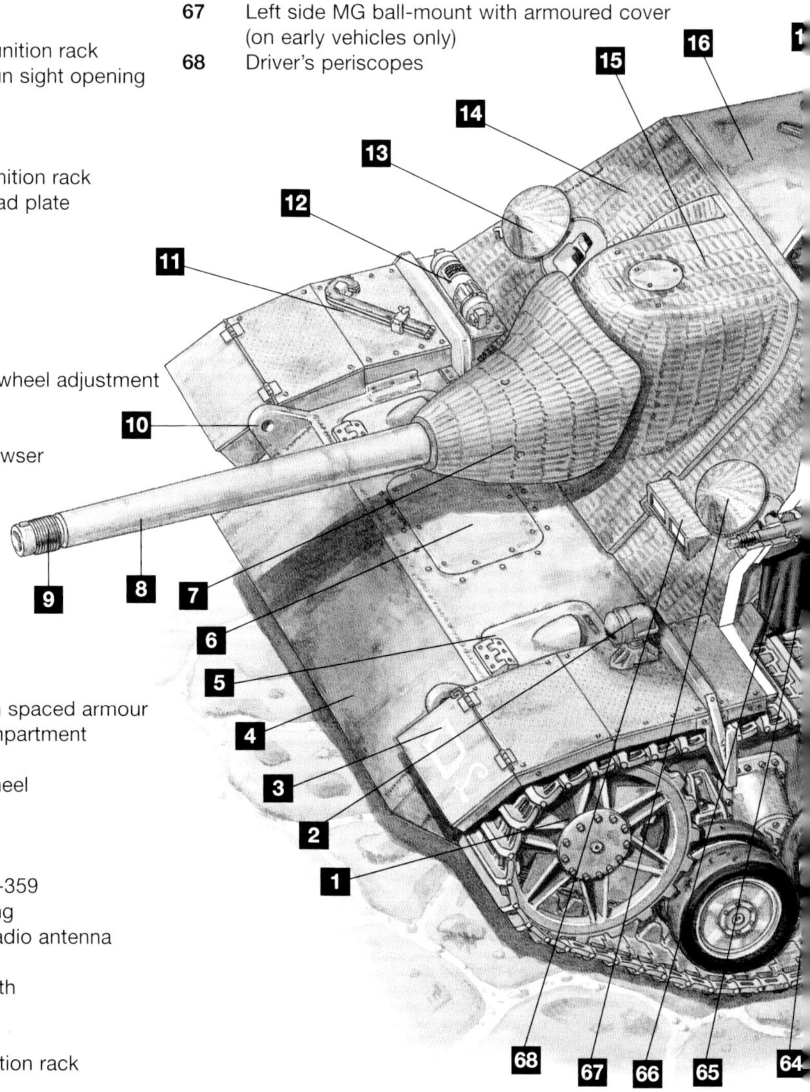

SPECIFICATION

Combat weight: 24,000kg
Motor: 11.867 ltr Maybach HL 120 TRM
265 metric horse power at 2,600 rpm
Power-to-weight ratio: 11.0 metric hp/ton
Ground pressure: 0.86kg/cm2
Overall length: 6,960mm
Width: 3,170mm
Transmission: 6 forward, 1 reverse
ZF SSG 76
Maximum speed (road): 40km/h
Maximum speed (cross country): 15km/h
Fuel capacity: 470 ltr
Maximum range: 210km at cruising speed
Fording depth: 1.00m
Armament: 7.5cm PaK 39/L48
Main gun ammunition: 79 rounds
50% 7.5cm Pzgr. (armour piercing)
50% 7.5cm Sprgr. (high explosive)
Muzzle velocity: 1,250 ft/s
Sight: Selbstfahrlafetten-Zielfehrnrohr 1a
(monocular, magnification x 2.5)
Secondary armament: 7.92mm MG42
MP44 or MP40
Ammunition: 1,200 rounds MG42;
384 MP

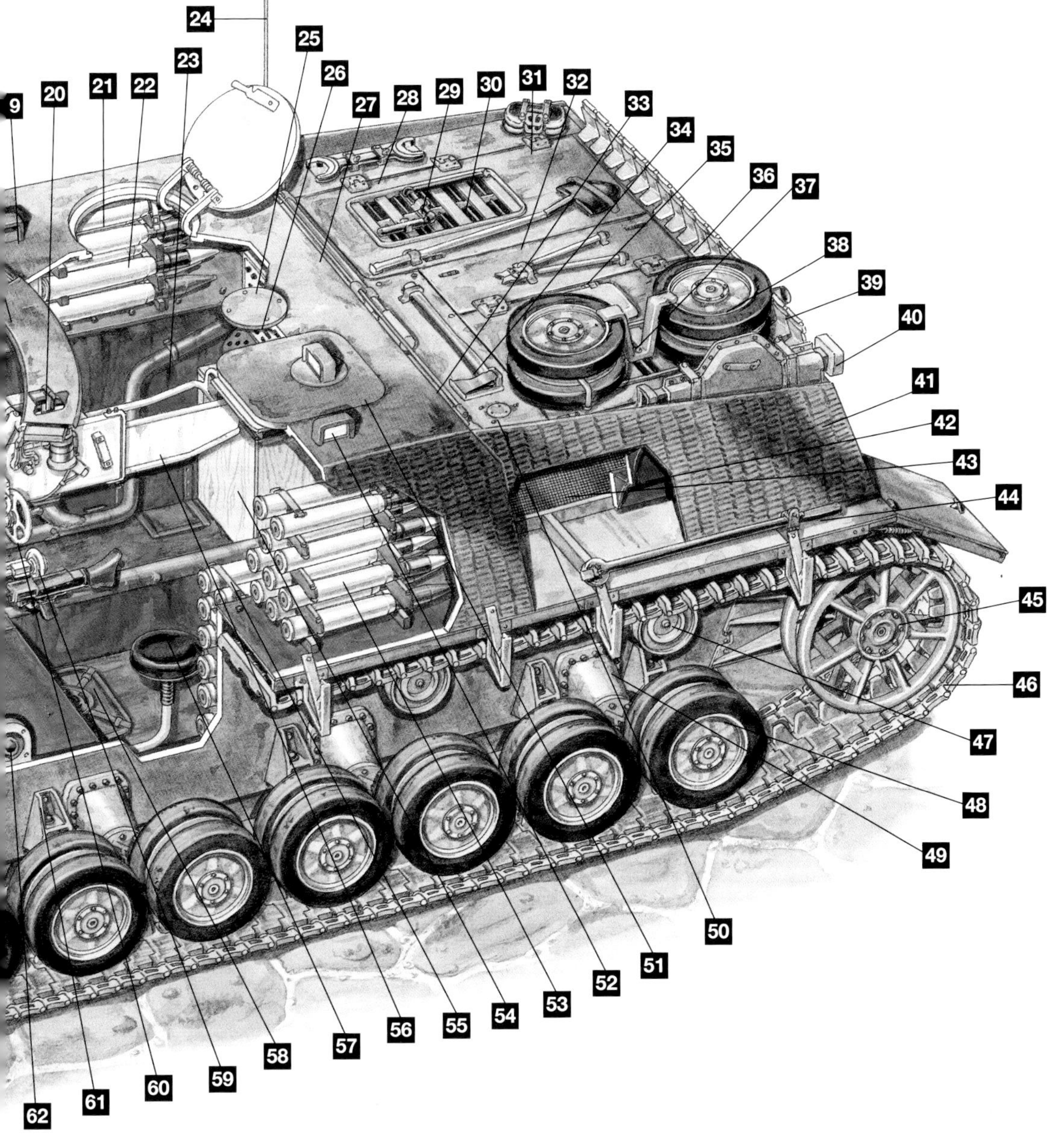

StuG III insignia
(see plate commentary for details)

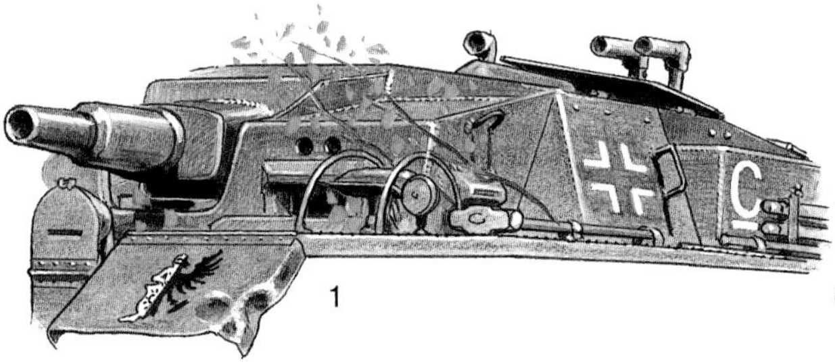

1

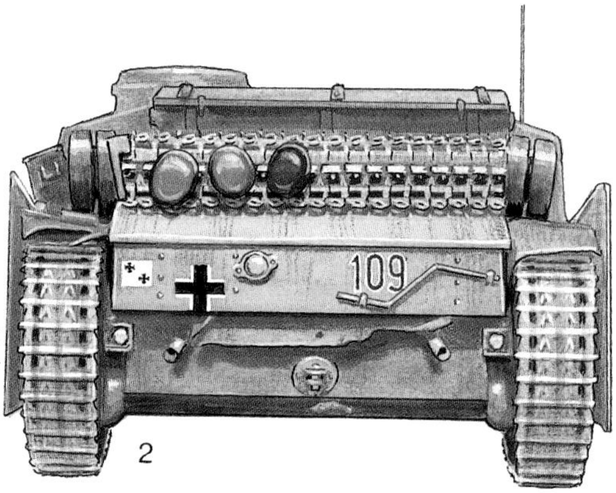

2

3

4

5

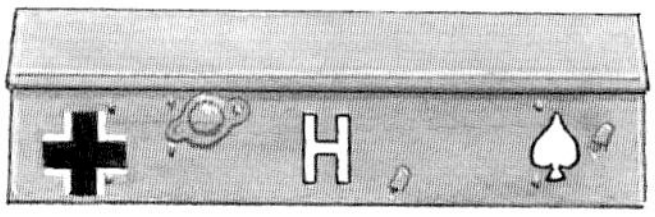

6

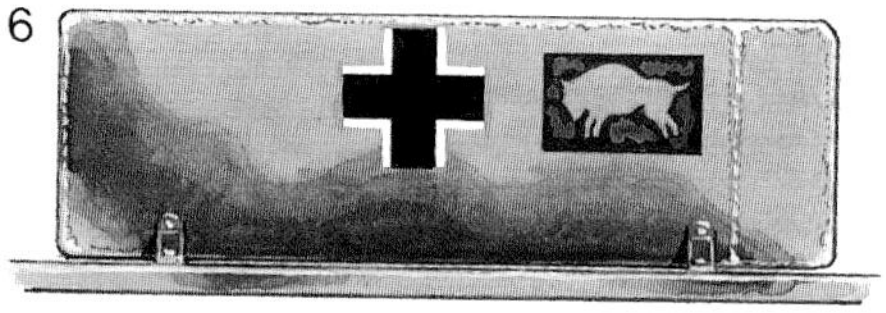

7

8

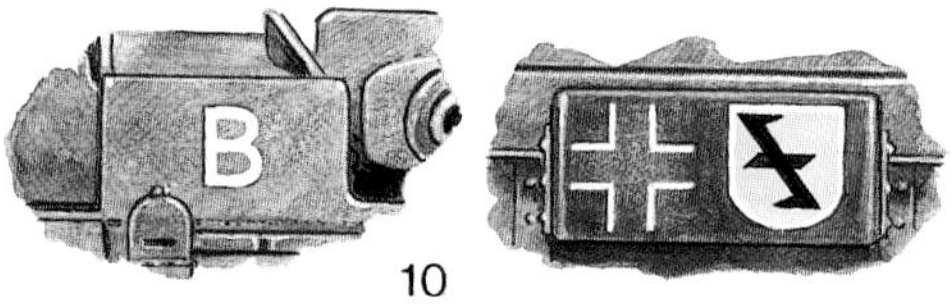

10

9

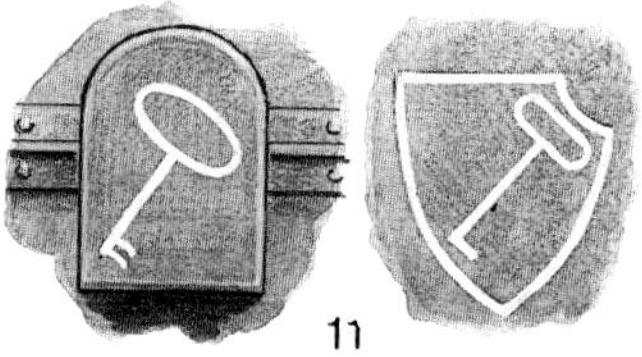

11

12

E

1. SdKfz 132 Marder II, 7.62cm Pak(r); 5.SS-Panzer-Grenadier-Division 'Wiking'; Orel sector, Russia, autumn 1943

2. SdKfz 131 Marder, 7.5cm Pak 40/2; 29.Infanterie-Division (Mot.); Russia, 1942–43

3. SdKfz 138 Marder III, 7.5cm Pak 40/3; unidentified army Panzerjäger unit, Russia, summer 1943

4. SdKfz 139 Marder III, 7.62cm Pak 36(r); 'Heidi' of unidentified unit, Russia, summer 1943

5. SdKfz 138 Marder, 7.5cm Pak 40/3 in rear housing; unidentified unit, Russia, autumn 1943

F

1. Brummbär, unit unknown; Russia, 1943. Inset: driver's position.

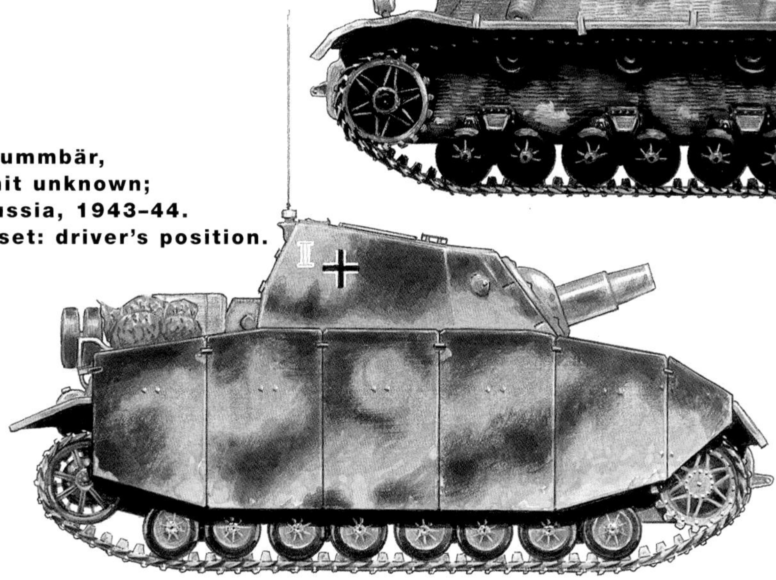

2. Brummbär, unit unknown; Russia, 1943–44. Inset: driver's position.

3. Brummbär, unit unknown; Italy, 1943–44.

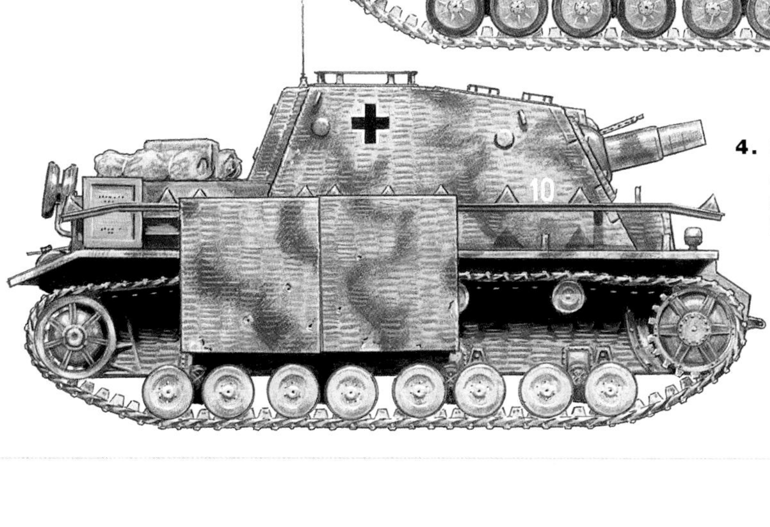

4. Brummbär Abschlusserie, 24.Panzer-Division; Russia, September 1944.

G

overall height being only 6ft 1in. The first gun to be fitted was the 75mm L48 Pak 39 with muzzle brake, later replaced by the more powerful 75mm L70 KwK 42, both weapons entering the vehicle via a 'Pig's Head' mantlet, to the right of which a conical hatch concealed a machine-gun. An interim model with a higher superstructure also mounted the 75mm L70, which had a muzzle velocity of 3,400ft/sec.

Like the assault gun, the Jagdpanzer IV had a crew of four, and its front armour was 80mm thick. It weighed 24 tons and had a speed of 24mph. Side skirts were often worn, and a coating of Zimmerit applied. It began entering service towards the end of 1943, gradually replacing the Marders in the tank destroyer battalions of the Panzer divisions. Although an efficient design, only 1,531 vehicles of this type had been constructed by the end of the war.

For all that it was a formidable opponent, the Jagdpanzer could be a mechanical mine of trouble after 900km running, and after 1,000km it required a complete overhaul. (The Tank Museum)

At the conference attended by Guderian following his appointment as Inspector General of Armoured Troops, he had stressed the need for a light assault gun with which to re-arm the infantry divisions' tank destroyer element. This appeared in 1944 under the title of Hetzer. (This is sometimes translated as 'Baiter'; however, the term 'Agitator' or 'Troublemaker' is more accurate.)

The Hetzer's angled superstructure occupied the whole of the 38t chassis on which it was mounted, the vehicle being armed with the 75mm L48 Pak 39 offset to the right of the fighting compartment, on the roof of which was a machine-gun that could be operated from inside the vehicle. The Hetzer was some 10 inches taller than the Jagdpanzer IV, but weighed eight tons less and was marginally faster. Its frontal armour was 60mm thick.

While much admired for its combination of hitting power and fine lines, the Hetzer was considered to be something of a pain in the neck by its crews. The driver, being located at the left front of the vehicle, suffered least from the cramped internal layout, but was badly placed if it came to leaving the vehicle in a hurry. Behind him sat the gunner, with the loader bringing up the rear, on the left of a gun designed to be loaded from the right and out of easy reach of his ammunition racks. The commander, seated at the right rear of the fighting compartment, was isolated from the others, and his optical aids were limited to periscopic binoculars which had to be used through an open hatch. Another effect of offsetting the gun to the right was an imbalance of traverse, so that while 11 degrees were obtainable to the right, the juxtaposition of breech and side armour permitted only 5 degrees to the left. In mitigation, it must be mentioned that offsetting was the only way in which the gun, four men and 41 large rounds of ammunition could be jammed into the available space. In spite of its unpopularity the Hetzer performed well, and continued to serve with the Swiss Army after the war.

The year 1944 also saw the introduction of the 45-ton Jagdpanther, armed with an 88mm L71 Pak 43/3 gun mounted in a well-sloped superstructure on a Panther tank chassis. Almost everything about the vehicle was right for a big-gun tank destroyer. The 80mm front armour was angled back to provide the equivalent of 160mm; the gun could penetrate every British, American and Russian tank in service; it stowed an adequate 60 rounds of ammunition; its weight was not excessive; its speed of 28mph was an asset; and the glacis contained a machine-gun for local defence. Only its height of 8ft 11in and the inevitable limited traverse, were against it. This efficient and beautifully proportioned gun had a crew of five, and served with the heavy tank destroyer battalions, replacing the vulnerable Nashorn. It had been hoped that 150 of these powerful vehicles could be built each month, but bombing and material shortages reduced the total eventual production run to 382.

'Guderian's Duck' – the Jagdpanzer IV, armed with the 75mm Pak 39, 48 calibres long. Marking positions and styles are typical of this vehicle. Side skirts were normally carried; note also the Zimmerit finish. (RAC Tank Museum)

Last in the line was the unwieldy Jagdtiger, which, although the most heavily armed AFV of the war, is best seen as an unnecessary waste of resources. Employing the complicated Tiger B tank chassis, it carried a 128mm L55 Pak 80 gun, protected by 250mm front, 80mm side and 40mm roof armour. The result stood at 9ft 3in high and weighed over 70 tons. While its speed of 23mph was most respectable for such a large vehicle, the sheer weight made attempts to tow 'like with like' merely result in a second breakdown.

38 rounds of main armament ammunition could be stowed, the heavy rounds being split for ease of handling by the loader, the vehicle's rate of fire thus being slower than that of tank destroyers using fixed ammunition. A machine-gun was housed in the glacis plate. Only a few of these monsters were constructed, serving with the heavy tank destroyer battalions and in support of SS Panzer formations.

Guderian's decision that the assault gun and the tank destroyer would gradually merge their roles was implemented throughout the introduction of this second generation of tank destroyers, a large proportion of the crews being drawn from the Assault Artillery.

TANK DESTROYER TACTICS

In spite of their aggressive title, it was not the function of tank destroyers to go out looking for tanks or to hunt them down in open combat. The tank destroyer was essentially a defensive concept, designed to destroy enemy armour by massed direct gunfire from carefully selected positions. These positions preferably lay on the flanks of the enemy's axis of advance or of any penetration he had made: ideally, they would offer cover from view and be located behind a tank obstacle, such as a river, marsh or minefield. The skill of the tank destroyer commander lay in his choosing several such positions *before* they were required, and then getting his guns into them in sufficient time. Concentration was the essence of good tank destroyer tactics, as it was in the employment of assault guns: dispersion into small units and single guns was bad practice.

Tank destroyer units serving with infantry formations often remained on a captured objective until it had been consolidated against counter-attack, and then moved back into reserve. If the infantry were attacked, the tank destroyers would be committed where the danger from enemy armour was greatest, adding their fire to that of the emplaced anti-tank guns. During a withdrawal, their mobility made them an essential part of the rearguard, enabling them to retire through a succession of pre-selected defensive positions into the new front line.

When working with Panzer divisions, the tank destroyers would form a firm fire base during the advance, as well as covering the flanks of the attack. If the tanks were successful, the tank destroyers would move up and form a fresh fire base for the next phase of the attack. If the tanks were not successful, or wished to 'write down' the enemy armour by a feigned retreat, they would retire through the tank destroyers' gun line, so leading their opponents into a fire trap. There were numerous instances of this both in North Africa and in Russia, British and Russian armoured formations paying dearly to learn the lesson that direct pursuit of an apparently beaten foe was not always a sound idea.

By virtue of their construction, tank destroyers generally carried a more powerful weapon than their opponents and engaged at the longest range possible, preferably beyond that at which the tanks' guns ceased to be effective. If, as frequently occurred on the open steppe or in the desert, the terrain did not offer suitable static fire positions, a running fight might develop, in which the tank destroyers used their mobility to keep the range open.

THE KEY TO MOBILITY

Mountings based on PzKpfw I chassis

The suspension consisted of four bogie wheels and a trailing idler, braced by an external beam and secured to the hull by bolts and quarter-elliptical leaf springs: track adjustment was obtained by altering the position of the idler. The power unit was a four-cylinder horizontally opposed air-cooled Krupp petrol engine which produced 57hp at 2,500rpm, fitted with one Solex downdraft carburettor for each bank, and an acceleration pump. Fuel capacity was 20 gallons, housed in two tanks mounted in the rear corners of the engine compartment. From the engine the drive passed through a two-plate dry clutch to the gearbox, which provided one reverse and five forward gears, and thence across the front of the vehicle to the drive sprockets. A clutch and brake steering system was employed with cooling supplied by a small fan. Direction was controlled by steering levers, each of which had two hand-grips, one for normal steering and the other with a

Allied air superiority in Normandy was total, and few movements were made by day: this Jagdpanzer IV risks a short hop, well camouflaged with a mass of foliage to break up its outline and cover reflecting surfaces. A second crew seems to be riding this gun. (Bundesarchiv)

Later models of the Jagdpanzer IV mounted a 70-cal. gun without muzzle brake. This badly battered example fell to two well-placed flank shots in a Normandy orchard. To the left of the mantlet is the swivelling conical cover of the machine-gun port. At 6ft 1in, three inches lower even than the early model StuG III, this formidable tank destroyer could adopt concealed firing positions in the slightest fold of dead ground. (RAC Tank Museum)

thumb plunger to act as a parking brake. Maximum recommended speed in gears was; 1st – 3mph; 2nd – 7mph; 3rd – 12mph; 4th – 20mph; 5th – 26mph.

Mountings based on PzKpfw II chassis

The PzKpfw II Ausf. A, B, C and F employed a five-wheel quarter-elliptical leaf spring suspension: the Ausf. D and E employed a torsion bar suspension mounting four large roadwheels, but lacked return rollers. Power was provided by a six-cylinder Maybach HL 62 petrol engine. The line of drive passed forward through a plate clutch to a crash gearbox which provided one reverse and six forward gears, then across the vehicle's front through reduction gears to the drive sprockets. Steering was effected by levers operating a clutch and brake mechanism. The driver's instruments included a speedometer, engine revolution counter, water temperature and oil pressure gauges. An electric starter was fitted but in low temperatures an inertia starter, operated through the stern plate, was generally used. Maximum recommended speed in gears was; 1st – 3mph; 2nd – 6mph; 3rd – 7mph; 4th – 13mph; 5th – 18mph; 6th – 25mph.

Mountings based on PzKpfw 38(t) chassis

The PzKpfw 38(t)'s suspension consisted of four large roadwheels per side, hanging in pairs from horizontal leaf springs bolted to the hull. Track tension could be adjusted by altering the position of the rear idler through hatches in the stern plate. The vehicle was driven by a 125hp Praga EPA water-cooled petrol engine which produced a maximum speed of 26mph: on later models this was fitted with twin carburretors which raised the output to 150hp and the maximum speed to 30mph. Power reached the front drive sprockets through a five forward/one reverse Praga-Wilson gearbox. A planetary steering system was employed, this being similar to that used by the PzKpfw 35(t), although it lacked the latter's pneumatic assistance. An electric starter was provided, as was an inertia starter operated from the rear of the vehicle: in emergencies the engine could be started manually from within the fighting compartment by a device located on the flywheel housing. Fuel capacity was 40 gallons, housed in two double-skin tanks mounted on either side of the engine compartment.

Mountings based on PzKpfw III chassis

The early development history of this chassis was plagued by difficulties with the suspension. These were solved in 1939 and resulted in the appearance of the Ausf. E, which employed a robust torsion bar system in conjunction with six small roadwheels per side. This model was powered by a 300hp 12-cylinder Maybach HL 120 TR engine, used in conjunction with a Maybach Variorex pre-selector gearbox with one reverse and 10 forward gears. In 1940 the latter was replaced in the Ausf.

H by the simpler Aphon synchromesh gearbox with one reverse and six forward gears. Maximum speed was approximately 25mph. The Ausf. H also incorporated a strengthened torsion bar suspension and track width was increased from 36cm to 40cm. To compensate for this new sprockets and idlers were introduced , although existing stocks were also used in conjunction with spacer rings. The line of drive ran through the gearbox and then across the front of the driving compartment to the drive sprockets. The final drive and steering brake assembly was complex and incorporated what was considered to be an excessive number of ball races; the track brake drums were provided with an air cooling system. The user handbook recommended a maximum engine speed of 2,600rpm for normal usage, but in hot climates, which included southern Russia as well as North Africa, it was suggested that employment of a lower gear than was necessary would produce cooler running. The principal components in the cooling system were two radiators through which air was dragged by two fans. The tractive effort of the upper reaches of both the Variorex and Aphon gearboxes was disappointing and it was recommended that 'when changing to a lower gear on turnings, hills or bad roads, two gears lower than the one already engaged should be selected.' The air filter was also considered to be inadequate in dusty climates, permitting sand to enter the lubrication system. An electric self-starter system was provided, but this was for use in emergencies only and never with a cold engine. Normal starting was by means of an inertia system, the two-man starting handle entering the engine compartment through the stern plate. Cold starting was assisted by a starter carburettor which could not be used in conjunction with the accelerator. Minimum unassisted working temperature for the engine was 50 degrees centigrade at 2,000rpm with an oil pressure of not less than 60lb per square inch.

The Hetzer – 'Troublemaker' – was another tank destroyer based on the old Czech 38t chassis. Note the pattern of the colour scheme; the remote control machine-gun on the roof; and, at the right rear corner of the roof, the commander's periscopic binoculars protruding through the horizontal front portion of the commander's hatch. (RAC Tank Museum)

Mountings based on PzKpfw IV chassis

While the PzKpfw IV was in its design stage, the army, against its better judgement, accepted a suspension system consisting of leaf-spring double bogie units. Bolted to the hull, each unit incorporated two small roadwheels, giving a total of eight per side. The increase in weight resulting from up-armouring later models overloaded the suspension, producing a tendency for the vehicle to yaw from side to side when no positive steering was applied: against this, replacement of complete bogie units, damaged in action, was a fairly simple matter. Track adjustment was achieved by movement of the rear idler, the mounting of which included an eccentric axle with a splined end. The axle could be rotated by a suitable tool and the adjustment retained by ratchet rings. On the Eastern Front track extensions known as

Ostketten were fitted to increase traction during the winter months. From Ausf. B onwards, the Maybach HL 120 TRM engine was installed, together with a new gearbox incorporating one reverse and six forward gears, producing much the same performance and limitations encountered with the PzKpfw III chassis. The principal components of the cooling system were two radiators coupled and mounted side by side at 25 degrees to the horizontal. Through these air was dragged by two fans driven by triple V-belts from the crankshaft, and the coolant circulated by a centrifugal water pump. Air was drawn into the engine compartment through a protected opening in the right-hand side of the hull and expelled through a similar opening on the left, provision also being made for aspiration into the fighting compartment, where the increased pressure assisted in the dispersal of fumes. The final drive and clutch-and-brake steering assembly were needlessly complex. Cooling for these units was provided by a centrifugal fan located on the left of the main clutch housing. Simultaneous application of the steering levers provided an effective parking brake. The engine was equipped with a 24-volt electric self-starter, and because an auxiliary generator kept the batteries fully charged this could be used more consistently than that fitted to the PzKpfw III. If the self-starter failed or extreme cold rendered its use inadvisable, an inertia starter, activated through the stern plate, would be used. This could be assisted by a starter carburettor and the driver depressing his clutch to reduce the drag of the gearbox oil. A further aid to cold starting on the Eastern Front was the Kuhlwasserubertragung or cold water exchanger. When one vehicle had been started and reached its normal operating temperature, the warm coolant was pumped to the next vehicle by the exchanger in return for cold coolant. In due course the rise in temperature would permit the second vehicle to be started, and so on.

A blown-out Hetzer, evidently caught in a village under bombardment. From the massive fractures in the main armour, it seems to have suffered a direct hit from a large-calibre high explosive shell. (RAC Tank Museum)

Mountings on the PzKpfw V Panther chassis

The Panther employed a torsion bar suspension system, with the difference that the bars were bent back on themselves in the manner of a hairpin and secured to the same side of the hull as the bogie unit they were supporting. The eight bogies on each side contained 16 interleaved large-diameter roadwheels. The system produced a good cross-country performance at the expense of adding to the vehicle's height and the compaction of mud, shingle and ice between the roadwheels, together with a degree of inaccessibilty. The track was tensioned by adjusting the position of the rear idler by means of an adjusting shaft accessible through a cover in the stern plate. The V-12 Maybach HL 230 P30 engine had a cylinder capacity of 23.88 ltr with a potential output of 700hp at 3,000rpm, although in service this was officially governed to 2,500rpm. The engine was water-cooled, two air-cooled radiators being located on each side of the engine compartment, linked by a compensating tank: some of the excess heat was ducted into the fighting compartment, giving a degree of comfort during the hard Eastern Front winters. In summer, however, the cooling system was initially unable to cope. Overheating, fuel vapourisation and engine fires resulted, leading to the installation of an automatic fire extinguisher system in the engine compartment. The drive passed from the engine to the three-plate dry clutch and thence to the seven-forward/one-reverse gearbox by means of two cardan shafts. The gearbox provided synchromesh engagement for all gears save first and reverse at 2,000–2,200rpm, but double de-clutching was required over 2,500rpm or when changing down at less than 1,500rpm: it was intolerant of heavy handling. The final drive assembly ran across the front of the vehicle and incorporated the same type of controlled differential steering employed by the British Churchill and the PzKpfw VI Tiger. 160 gallons of fuel were carried, contained in five inter-connected tanks, one against the rear wall of the engine compartment and two on each side. An electric self-starter was fitted, but if the vehicle's batteries were flat or cold temperatures prevailed a Bosch inertia starter could be activated through the stern plate. While the PzKpfw V chassis contained many excellent features, a prisoner taken in 1944 indicated that its users had serious reservations: 'The early models had a great deal of engine, clutch and gearbox trouble which was rapidly cured, but the remaining defect is the weakness of the reduction drives to the drive sprockets. These give constant trouble as they are under-tensioned for the torque they have to transmit. A Panther is regarded as a potential mine of trouble from 900kms onwards; by 1,000kms it will probably need a 100 per cent overhaul of the drive at least.'

Mounting on Tiger (P) chassis

This chassis was never adopted for tank use and the only vehicle to be produced on it was the Elefant (or Ferdinand) heavy tank destroyer. Unusually, a longitudinal torsion bar suspension was employed in conjunction with six roadwheels in bogie pairs per side and a rear drive sprocket. In action, the torsion bars proved unequal to the stresses imposed by the vertical movement of the roadwheels on rough terrain. The principal point of interest, however, lies in the use of a petrol-electric drive. The primary power unit consisted of twin Maybach HL 120 TRM water-cooled petrol engines producing 530hp at 2,600rpm. These were coupled to a generator from which power was transmitted to two electric motors and thence to the drive

The elegant Jagdpanther, seen here clean and with 'textbook' stowage during unit working-up exercises in Germany. The cylindrical auxiliary fuel tank carried on the superstructure side can be seen clearly. (Bundesarchiv)

sprockets. In itself, the system was complex and unreliable: furthermore, its electrical components required large quantities of copper at a time when that commodity was in critically short supply.

Mounting based on PzKpfw VI Tiger (E) chassis

The chassis was carried on an eight triple-bogie interleaved torsion bar suspension with a front drive sprocket. Track adjustment was obtained through the cranked mounting of the rear idler wheel, the draw bolts, by means of which the adjustment was made, being internally mounted but accessible from outside by the removal of a domed cover on each side of the stern plate. Originally all roadwheels were rubber-tyred but in 1944 all-steel resilient roadwheels wheels were introduced and the outer set were abandoned. While efficient, the interleaved wheel arrangement was vulnerable to compaction by mud, stones and ice. The 725mm service track rendered the vehicle too wide for rail transit and for this a 520mm track could be fitted after the outer wheel from each bogie had been removed. The vehicle was powered by a Maybach HL230 V-12 700hp water-cooled engine. Two linked radiators were located on each side of the engine compartment and through these air was drawn by four fans driven by the timing gears, then expelled through louvres in the engine deck. Two petrol tanks were also fitted at each side of the engine compartment, coupled together and holding a total of 125 gallons. Petrol consumption averaged 2.75 gallons per mile across country. From the engine the drive passed to a hydraulically operated pre-selector Olvar gearbox giving eight forward and four

Jagdpanthers moving up to the Invasion Front, 1944. Spaces are left in the Zimmerit to the right of the drivers' visors, for attaching rail shipping labels. The performance of this tank destroyer did not belie its beautiful lines and purposeful air: in less than two minutes three of them all but destroyed a squadron of Churchill tanks on 30 July at Hill 226. (Bundesarchiv)

Weighing over 70 tons and mounting a 128mm gun, the Jagdtiger was the most powerful armoured fighting vehicle of the war. Its practical value was less striking however: it was subject to breakdowns and had a low rate of fire. (RAC Tank Museum)

reverse gears, used in conjunction with three hydraulic cylinder selectors, thus providing the widest possible range of flexibility. Beyond the gearbox the drive passed across the front of the vehicle through an extremely complex final-drive mechanism to the drive sprockets. A hydraulic regenerative controlled-differential steering system was employed, similar to the Merritt-Brown system installed in the British Churchill, the driver using his steering wheel to impose different speeds on the sun-wheels of the epicyclic gearing. While over-complex and absorbing construction time and resources that could have been put to better use elsewhere, these systems made the vehicle a pleasure to drive. If the power steering failed, the driver could fall back on two steering levers which acted on the vehicle brakes, but these were intended for emergency use only. A 24-volt self-starter motor was fitted, as was a Bosch inertia starter which was operated by a handle inserted through the stern plate: in low temperatures a starter carburettor could be employed.

Mounting on PzKpfw VI Tiger B chassis

The automotive layout and power train of the Tiger Ausf. B were very similar to that of the Ausf. E but incorporated a number of improvements. For example, the vehicle travelled on nine double-bogie units per side, the roadwheels being overlapped rather than interleaved. Fuel capacity was increased to 175 gallons and an enlarged water cooling system installed. In addition to the electric and inertia starters, the Ausf. B could also employ an emergency petrol starting engine. This was positioned on two brackets close to the rear end of the crankshaft, which was engaged by dogs. As with the Ausf. E, the vehicle was provided with a narrower set of tracks for transit by rail, but as the bogie units were narrower the track changing operation was simplified by not having to remove the outer wheels.

ENDNOTE

The assault gun and the big-gun tank destroyer were phenomena of World War 2, and disappeared because technology had rendered them obsolete. The establishment of all-arms battlegroups, including a multitude of armoured personnel carriers, has meant that the infantry no longer require specialist vehicles to shoot them into an objective, the role of the assault gun being absorbed into that of the tank. The big-gun tank destroyer survived only so long as it could outrange the tank. Once tanks began carrying larger guns, the need for it disappeared. Today's tank destroyers carry guided missiles, a far cry from the simple principles of the early Marders. In their day both weapon systems offered ingenious, inexpensive and effective solutions to specific battlefield problems.

Assault Gun Battery 1940

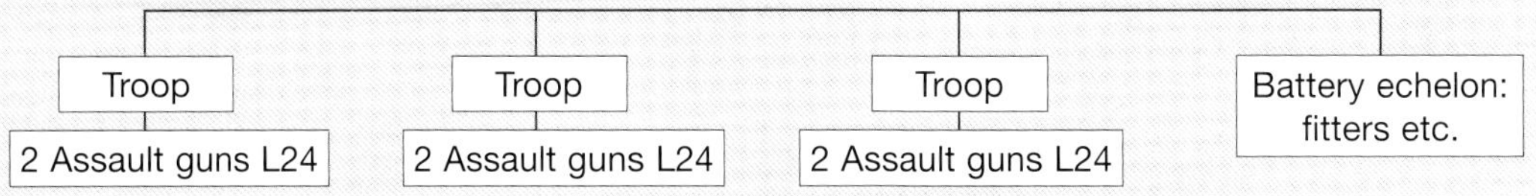

Assault Gun Battalion 1941

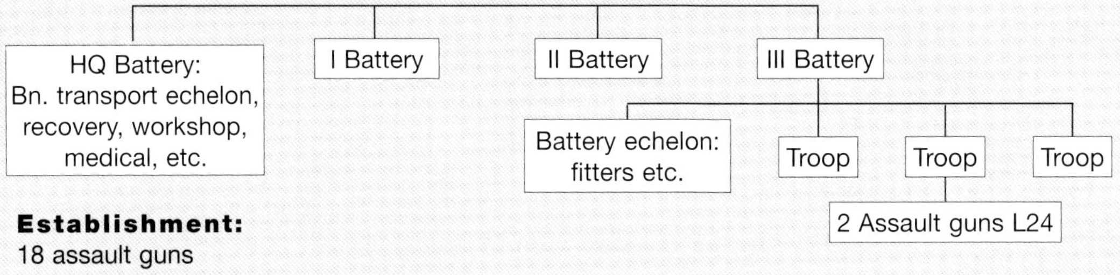

INTERIM ORDER OF BATTLE Assault Gun Battalion, 1942

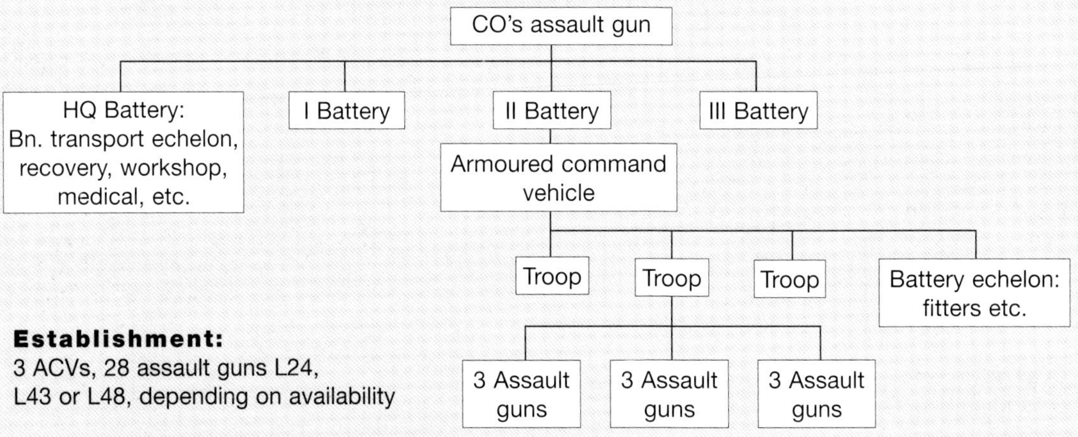

Assault Gun Battalion (Brigade) 1942–45

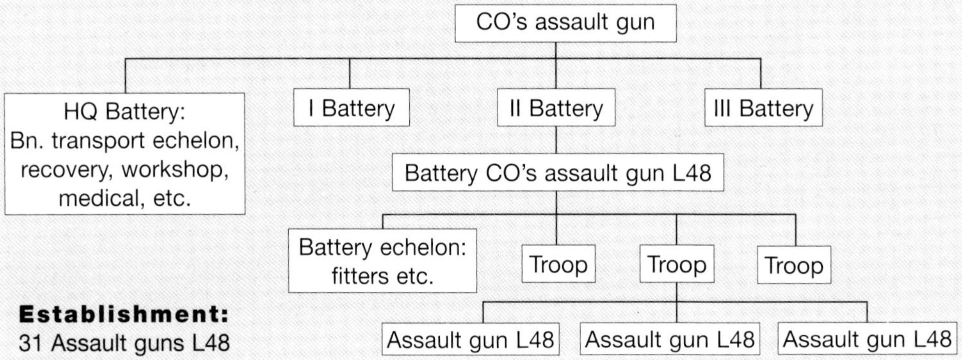

Assault Artillery Brigade 1944–45

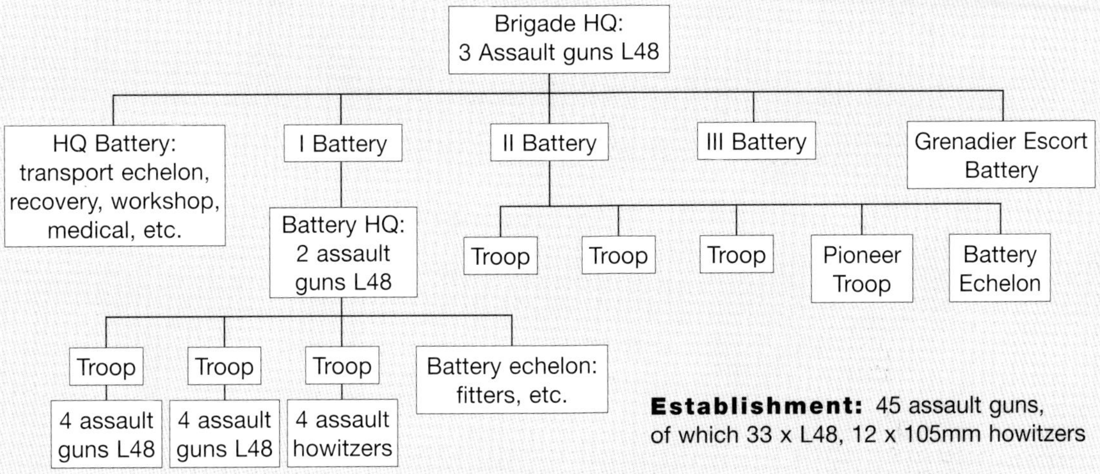

Establishment: 45 assault guns, of which 33 x L48, 12 x 105mm howitzers

ORDER OF BATTLE
653 and 654 Heavy Tank Destroyer Battalions; Kursk, July 1943

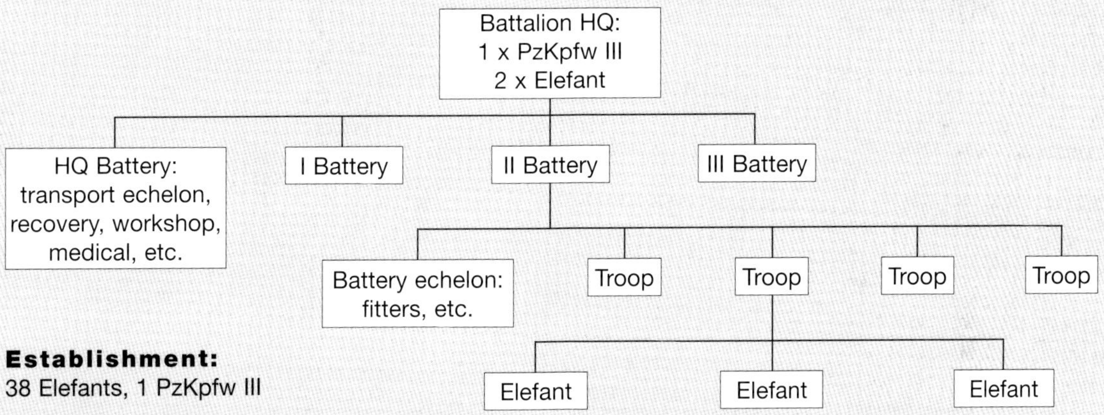

Establishment:
38 Elefants, 1 PzKpfw III

ORDER OF BATTLE
Heavy Tank Destroyer Battalion

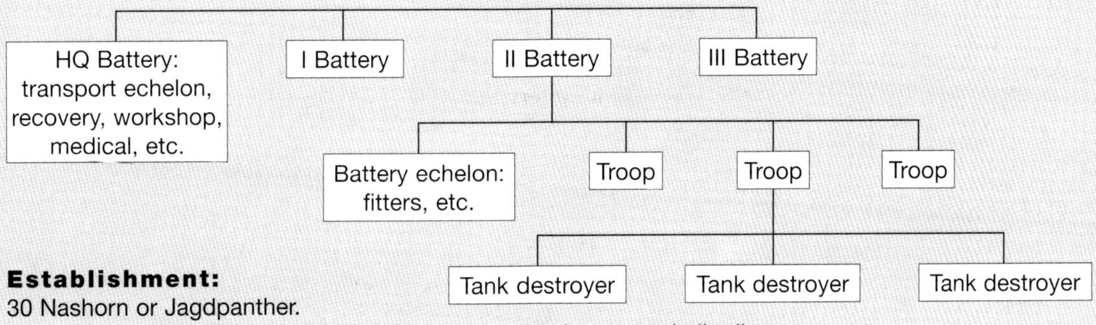

Establishment:
30 Nashorn or Jagdpanther.
The standard Tank Destroyer Battalion was organised on very similar lines.

THE PLATES

The colour plates in the centre of this book are based on wartime photos of specific individual vehicles, and do not represent theoretical reconstructions from written data. In some cases it has been necessary to make an educated guess at camouflage colours: in black and white photos it is extremely hard to tell the difference between the dark green and dark red-brown paints issued to vehicle crews from 1943 onwards, for camouflaging the basic ochre yellow factory finish of their equipment. Different degrees of dilution, and the effects of exposure over different periods, make this inevitable. Sometimes the setting of a photograph is helpful, as in Plate G3, where the photo from which we worked shows a vehicle concealed in an Italian grove in summertime. Here it is a fair assumption that green rather than brown was used for the camouflage.

A: StuG III Ausf. G fighting compartment interiors

Top This is a command vehicle, with radio equipment on both sides of the fighting compartment – the normal commander's set may just be seen above his arms. The shelf on which this rested seems from photos to have been enclosed along the front, in some vehicles, by a 'fiddle' made from a plank, and loose gear was stowed along it. The Leutnant battery commander, his silver-piped

According to Churchill crewmen who were present, this Nashorn (Rhino) caused a great deal of trouble before it was stalked and killed with a PIAT by Canadian infantry. The action took place during the storming of the Hitler Line, Italy, 26 May 1944. (Public Archives of Canada)

Einheitsfeldmütze turned backwards in the style of a U-boat commander, is using the binocular sight protruding from his cupola. The Unteroffizzier gunner uses his own sighting periscope, which protrudes through an opening in the compartment roof. He and the loader, partly visible on the right, wear the earlier Feldmützen. The latter loads a round of HE, its olive green head stencilled 4 13 1 in black. Below the ventilator intake and trunking on the rear wall is a rack and a row of spring clips for stick grenades; above it, an MP40 and a set of magazine pouches are fixed for quick access.

Bottom The loader in the bottom view wears the reversible padded winter jacket, 'snow' side out, and the 1943 Einheitsfeldmütze. A second MP40 and its magazine pouches are fixed to the rear wall behind him. On his right is the extra command radio equipment which he operated; extra ammunition is piled in front of this. The whole crew wear standard headset and throat-microphone equipment. Spare sets seem often to have been hung on the rear bulkhead. Loose personal gear would normally be packed in every available space, such as along the top of the radios.

B1: StuG III Ausf. B, 2 Batterie, StuG Abteilung 192; Gomel airfield, Russia, August 1941

A heavily stowed vehicle; in the photo on which this painting is based, No. 25 is more heavily camouflaged with birch saplings – we have removed most of them for clarity. The death's-head insignia is that of the Abteilung and seems to have been carried by all vehicles; other photos show it repeated (alternatively marked?) on the left hull side (both sides?) forward of the radio housing. The letter 'S' was carried by the Batteriestab (HQ) guns; in this case 25 seems to have been removed from its usual place in the order of battle to do temporary duty as the battery commander's mount, it being a simple matter to paint the relevant insignia in the standard position. The tactical marking stencilled on the bow plate (detail inset) appears from comparison of tones to be in yellow. The dust of the summer advance of 1941 seems to have given the overall 'Panzer grey' finish a very faded appearance, which we have not tried to reproduce.

The unditching beam appears in many photos: sometimes an untrimmed log was used. Typical exterior stowage included jerrycans, wooden crates, folded tarpaulins, crew helmets and canteens, etc. The crew wore the field grey uniform of Panzer cut, with dark green collar patches trimmed red, and silver skulls. Although it was not generally seen at the front, the grey version of the beret-like Schutzmütze is clearly visible in a published photo of an NCO of this battalion at this period.

B2: StuG III Ausf. B, 3 Batterie, StuG Abteilung 203; Smolensk, Russia, July 1941

The elephant insignia of the Abteilung appears on this vehicle in both left and right side presentations, on hull side and rear respectively. The white '33' identifies the third gun of the third battery. Again, the overall finish is 'Panzer grey'. Improvised jerrycan racks of iron strip were often welded to the rear deck, and took many forms. Photos of assault guns in Russia throughout the war seem to indicate that whole spare sets of tracks and many spare bogies were often carried on the rear deck, the former coiled loosely on their sides. A bucket was a typical item of 'acquired' external stowage. The elephant and two-digit marking style seems to have been retained by this battalion well into the war – see Plate E, item 7.

C1: StuG III Ausf. G, 2.SS-Panzer-Division 'Das Reich'; Kursk, Russia, July 1943

The most striking structural difference from the early versions, seen from this angle, is the adoption of spaced skirt-armour hung from side rails. The vehicle is now finished in overall dark ochre yellow: the crew added the brown and green camouflage paint at their own or their commanders' discretion, and patterns were enormously varied. A loose, random series of 'curls' is visible in the photo from which we take this illustration. It also shows an air recognition flag tied over the rear stowage; an absence of any unit/vehicle identification numbers (not uncommon, from 1943 onwards); and the tactical, national and divisional insignia all marked on the rear hull plate. The tactical insignia identifies this gun as of I Batterie, SS-Sturmgeschütz-Abteilung 2. The divisional insignia is the two-bar type adopted temporarily for the Kursk operation: see also Plate E, item 12. This is a command vehicle, as witnessed by the two radio aerials.

C2: StuG III Ausf. G, 16.SS-Panzer-Grenadier-Division 'Reichsführer'; Italy, January 1944

One of the less well known Waffen-SS formations, this understrength division did provide some units for the Anzio fighting, and several photos of StuG IIIs in transit in Italy have been published. The vehicle on which we base this illustration was apparently finished in factory ochre all over, without camouflage: note also that the commander wears the normal four-pocket Waffen-SS service tunic instead of the special double-breasted cross-over jacket of the vehicle crews' uniform. The individual vehicle number is marked on the hull side in black, and the tactical marking on the bow plate, identifying 1 Batterie of the Abteilung, shows a slight variation from the example above. One photo which shows the rear of a column of these vehicles shows the nonstandard presentation of the national cross illustrated in the upper inset detail. The black SS rune insignia marked on the mantlet and rear plate of all vehicles in this sequence of photos was not, in fact, the official divisional insignia, and is not fully explained: the recognised vehicle marking for 'Reichsführer' was a stylised presentation of Heinrich Himmler's oakleaf collar patch design.

D: Jagdpanzer IV (SdKfz 162), 3.Kompanie, Panzer Jäger Abteilung 130, Panzer Lehr Division; Normandy, 1944

The external vertical surfaces of the Jagdpanzer IV in this period were coated in Zimmerit anti-magnetic paste and the whole vehicle painted in dunkelgelb RAL 7028 (dark yellow) at the factory in Plauen before delivery. For camouflage, the crews were issued with 2kg cans of paste which could be diluted with any petroleum based liquid, or even water, and applied with a spray in broad stripes and patches on top of the base dunkelgelb. Two colours of paste were issued; dunkelgrün RAL 6003 (dark olive green) and rotbraun RAL 8017 (dark chocolate brown). However, in this case the crew only used the dunkelgrün overspray paste.

Most upper parts of the fighting compartment interior were painted in elfenbein RAL 1001 (ivory yellow) while lower parts, floors, etc. and the motor compartment were left in grundierfarbe rotbraun RAL 8012 (redprimer).

This is one of the 31 Jagdpanzer IV shipped to the Panzer Jäger Abteilung 130 of Panzer Lehr Division on the 17 March 1944 for use in the West. They opposed the Allied advance after D-day. The tactical numbers of the 3.Kompanie, Panzer Lehr Division were probably blue outlined in white. The unit markings are on the driver-side front mudflap.

E: Insignia details, StuG III

1. Hull and trackguard markings of a StuG III Ausf. D of 189 Sturmgeschütz Abteilung, photographed in Russia early in the 1941 campaign. The national cross is in white outline on the overall dark 'Panzer grey' finish: the white 'C' is presumably a battery letter. The left front trackguard bears the Abteilung insignia, a fairly uncommon placing. This is vertically halved: on the right, half of a black heraldic spread eagle, and on the left, in white with black details, half of a medieval knight standing with a grounded shield bearing a sunburst device. Note unusual headlamp guards; and periscopic binoculars protruding from the commander's hatch.

2. Rear view of a StuG III Ausf. G of 286 StuG Brigade, Russia 1943–44. The heavily stowed vehicle has skirt armour, and a large wooden stowage crate across the rear deck. The finish appears to be weathered but uncamouflaged overall dark ochre. The national cross appears in the position which seems to have been typical for the StuG III: left of it is the brigade insignia, two small black Maltese crosses on white, and at the right end of the rear hull plate is the individual vehicle number in red trimmed with white – presumably 9th vehicle, 1st Battery.

3. Rear hull markings of StuG III Ausf. G of 259 StuG Brigade, Russia 1943–44. The ochre factory finish appears to be very lightly and indistinctly camouflaged with either green or brown. The black individual vehicle number appears alone, without battery identification, and the brigade insignia is stencilled to the right in black and white – an eagle's head on a white Maltese cross on a black shield.

4. Hull right side markings of StuG III Ausf. F, Panzer-Grenadier-Division 'Großdeutschland', Russia, summer 1943. Soft brown camouflage streaks over ochre finish. Battery and vehicle numbers in plain white, and divisional insignia forward of cross.

5. Rear hull markings of an unidentified StuG III Ausf. G, Russia, autumn 1943. No side markings are apparent in the photograph of this vehicle. The ace of spades is not listed as a unit insignia in German records, but these are very incomplete.

6. Hull right side markings of StuG III Ausf. F, 191 StuG Brigade, Russia, 1943. The brigade's charging bison insignia appears to have been marked with a stencil when the original 'Panzer grey' overall finish was re-painted ochre.

7. Right hull side markings of StuG III Ausf. F of 203 StuG Brigade in Russia, 1943. Study of the relative tones in a photograph of this vehicle suggests that the brigade's elephant insignia appeared in the ochre ground colour on a black panel. Note the unusually hard-edged camouflage pattern, probably in green.

8. Hull side markings of StuG III Ausf. G of 8. SS-Kavallerie-Division 'Florian Geyer'; Russia, winter 1943–44. A most unusual, in fact in the editor's experience unique example of the actual name of an SS division being painted on a vehicle. The photograph from which it is taken shows the gun to be camouflaged in a dirty but complete coat of white, apart from the ochre patch left around the name. No other markings are apparent, on the rear hull plate at least. The vehicle is heavily stowed, with a large wooden crate on the engine decking, and carries a swastika air recognition flag over the rear deck stowage.

9. Mantlet decoration of StuG III Ausf. G of an unidentified unit, Russia, 1944. The photograph shows a complete battery, at least, all with the same decoration. Finish appears to be uncamouflaged ochre, and markings are limited, as far as we can tell, to a three-figure number carried centrally on the skirt armour, e.g. 'III', possibly in red trimmed with white.

10. Front and rear markings of StuG III Ausf. E of 2 Battery, StuG Abteilung 249, photographed in the Crimea in the summer of 1942. The overall grey finish is heavily weathered and dusty in the photographs. The 'B' could be either a battery or an individual vehicle identification – another vehicle in the same group bears the letter 'G', so the latter is perhaps more likely. The 'Wolfs Tongue' rune is the Abteilung insignia.

11. Styles of insignia used by the assault gun battery of the 'Leibstandarte SS Adolf Hitler' regimental combat group in France, 1940 (left); and by the StuG battalion of the enlarged 'LAH' Brigade in the Balkans, 1941 (right) – both in rear hull presentations. In the former case an additional, apparently battery, emblem has been photographed – a simple left profile of a dog's head (wolf?) in solid white, painted centrally on the rear plate. (StuG Abteilung 261 later used a wolf-head emblem.)

12. Divisional insignia carried by StuG III of 3.SS-Panzer-Division 'Totenkopf' during the Kursk offensive, July 1943.

F: Marder colour schemes

These illustrations display the full range of typical applications of the brown and green camouflage paints over the ochre finish; at the same time they emphasise the great diversity of chassis and armament types embraced by the 'Marder' SP gun series. In each case we have reproduced, roughly, the uniforms worn by personnel in the photos from which we take the vehicle subjects.

1. SdKfz 132 Marder II, 7.62cm Pak(r); 5.SS-Panzer-Grenadier-Division 'Wiking'; Orel sector, Russia, autumn 1943. Plain, uncamouflaged ochre, with light foliage camouflage added to wires strung around the external hull brackets. No painted markings apart from national cross. The chalked 'WK' on the hull side well forward may be some purely co-incidental temporary addition; but it is tempting to equate it with a hasty abbreviation of the divisional name? Crew in field grey service uniform and camouflage smocks, helmet covers.

2. SdKfz 131 Marder, 7.5cm Pak 40/2; 29.Infanterie-Division (Mot.); Russia, 1942–43. This is a very well-known and widely illustrated vehicle, 'Kohlenklau'. A soft, sparse blotching of much-diluted green paint has been added to the factory finish. Markings are lavish: apart from the cross and the cartoon 'coal-thief' above the name, there is a red triangle with a white centre bearing '1A' in black – presumably a troop and vehicle insignia, but untypical in the extreme. Photos of this vehicle have been identified as being taken in autumn 1942: they are certainly set in muddy enough terrain for the Russian autumn, and the division's virtual annihilation in the Stalingrad fighting that winter suggests a date of 1942 for the photos. If this is so, then the use of the Einheitsfeldmütze by the crew is puzzling, as it is not supposed to have been issued until the next year. It is worn with the normal four-pocket service uniform.

3. SdKfz 138 Marder III, 7.5cm Pak 40/3; unidentified army Panzerjäger unit, Russia, summer 1943. The green and brown paints have now been applied in large, irregular, soft-edged areas all over the upper surface of the vehicle, though not extending over the suspension. Batterie, Zug and individual gun numbers are marked in the normal Panzer style. Crewmen wear field grey uniforms of Panzer cut, with black, pink-piped collar patches bearing silver skulls, and pink-piped shoulder-straps.

Greek meets Greek, North Germany, 1945: a Jagdpanther and a 90mm M36, both knocked out, illustrate two contrasting approaches to tank destroyer design. (Capt. C.A. Heckethorn, 899 TD Bn.)

4. SdKfz 139 Marder III, 7.62cm Pak 36(r); 'Heidi' of an unidentified unit, Russia, summer 1943. The green and brown areas now cover the whole vehicle, leaving only thin channels of ochre visible between them. The suspension and lower hull side are camouflaged – even the wooden crate added at the rear for crew stowage has been roughly streaked with paint. Markings are limited to the cross, and a vehicle name in black shadowed with white – 'Heidei', painted just behind the aerial. Crew wear grey shirts, reed green denim trousers, field grey caps.

5. SdKfz 138 Marder, 7.5cm Pak 40/3 in rear housing; unidentified unit, Russia, autumn 1943. A pattern of smaller, harder-edged streaks of brown and green, used with little dilution and apparently scrubbed on with a brush or broom, completely covers all surfaces of this vehicle, leaving only the narrowest of ochre divisions. There are no markings visible apart from the cross. A tarpaulin is rigged over the fighting compartment – the photo on which we base this was taken in rainy weather. The crew seem to wear grey four-pocket service uniforms and Einheitsfeldmützen.

G: Brummbär colour schemes and markings

Details of unit deployment of this impressive vehicle are sparse. These illustrations cannot be tied down with any confidence except in one case, and thus are interpretations of wartime photographs.

1. and **2.** These are both apparently vehicles which served in Russia in 1943–44, and which were photographed after being disabled. **1.** has a coating of Zimmerit paste over all vertical surfaces, including the exposed hull sides above the road wheels – an unusual feature, but one which made sense for a vehicle designed for street fighting at short ranges, as this area was particularly vulnerable. The dark green and brown paint seems to have been added to the ochre factory finish by this crew in an unusually crude way. The white number presumably identifies a troop and the individual vehicle. Most of the few available photos show Brummbär crews wearing black Panzer uniform. **2.** has no Zimmerit, but has side skirt armour and a more conventional camouflage finish. The small Roman 'II' in yellow trimmed with white might suggest a 2nd Battalion HQ vehicle. The inset scrap views show in each case the driver's compartment of these vehicles – the early box-type with top periscope, and the later type using the same armoured visor arrangement as the Tiger Ausf. E tank.

3. appears in the photograph on page 10, taking cover in an Italian grove. The picture was clearly taken in summer, and given the season and terrain it is a fair assumption that the interesting 'creeper'-like camouflage spreading down from the top of the fighting compartment was in green rather than brown. Again, Zimmerit paste protects the hull and fighting compartment vertical surfaces from the attachment of magnetic or 'sticky' bombs – though not, apparently, the thin sheet metal of the rear stowage bins, which were unimportant. The single white '1' must be a vehicle number rather than a unit identification. **4.** is an example of the improved Brummbär Abschlusserie – note the added machine-gun in the ball mount high on the left side of the glacis, and the new protruding section in the rear face of the fighting compartment. The skirt armour now attaches by quick-release hooks and brackets, so that collisions with walls etc. no longer tear the plates but simply knock them off. The Zimmerit again extends right down among the suspension details, and the camouflage is entirely conventional. This vehicle has been identified in print as belonging to 1 Batterie, Panzer-Artillerie-Regiment 89, 24.Panzer-Division, serving in Russia in September 1944. The white '10' low and forward on the hull probably identifies a battery HQ vehicle, if this is the case. (Modellers might care to recall, if planning to reproduce this colour scheme, that 24.Panzer-Division personnel wore golden yellow instead of rose pink piping on their black Panzer uniforms, in memory of the old 1.Kavallerie-Division from which the formation was raised.)

An abandoned Jagdtiger in a German town, evidently a victim of mechanical breakdown. (Capt. C.A. Heckethorn, 899 TD Bn.)